Destructive Turf Insects

Second Edition

By

Dr. Harry D. Niemczyk

Professor Emeritus, Department of Entomology
The Ohio Agricultural and Development Center, The Ohio State University
Wooster, Ohio

&

Dr. David J. Shetlar

Associate Professor, Department of Entomology
The Ohio State University
Columbus, Ohio

H.D.N. Books | Wooster, Ohio

Copyright © 2000 by **Dr. Harry D. Niemczyk**

dba H.D.N. Books
2935 E. Smithville West Road
Wooster, Ohio 44691-1056

Printed in the United States of America.

ISBN 1-883751-14-4

Library of Congress Control Number: 00 - 134238

The Authors

Harry D. Niemczyk, Ph.D.

Dr. Harry D. Niemczyk is Emeritus Professor Turfgrass Entomology at the Ohio Agricultural Research and Development Center, The Ohio State University, Wooster, Ohio. A member of the faculty at OARDC/OSU since 1964 and Professor since 1971, Dr. Niemczyk received his B.S., M.S., and Ph.D. degrees from Michigan State University. His research on behavior and mobility of pesticides in turf and the biology, ecology, and control of insects and other arthropods in turfgrass has been published in scientific journals and frequently in trade publications. He was coeditor and author of the book, Advances in Turfgrass Entomology. He practices his profession as a private consultant to the turfgrass management industry, and develops training and insect control programs for lawn care firms, sod farms, and golf courses, and he presents seminars on turfgrass insects. He is a member of the Entomological Society of America, American Society of Agronomy, International Turfgrass Society and the Ohio Turfgrass Foundation. Dr. Niemczyk received the Professional Excellence, Man-of-the-Year, and Lifetime Membership awards from the Ohio Turfgrass Foundation. He serves as a member of the Technical Advisory Committee of the Ohio Turfgrass Foundation, is a member of the Education Staff of the Golf Course Superintendent's Association of America, and serves on the Board of Directors of the Musser International Turf Foundation.

David J. Shetlar, Ph.D.

Dr. David J. Shetlar is Associate Professor of Landscape Entomology at The Ohio State University. He performs turfgrass entomological research at The Ohio State University, Ohio Turfgrass Foundation, Turfgrass Research and Education Facility, with his extension and teaching efforts based in Columbus, Ohio. Dr. Shetlar received his B.S. and M.S. in zoology from the University of Oklahoma and his Ph.D. in entomology from Penn State. He was an assistant professor of entomology at Penn State from 1978 to 1983. He then joined the ChemLawn Research and Development Center in 1984 as Research Scientist, Turfgrass Entomology. Here he had responsibilities to develop new products and programs for management of turfgrass pests in North America. In 1990, he joined The Ohio State University. His research has centered on the ecology and behavior of bluegrass billbug, black cutworm and sod webworms. He is actively involved in developing control programs, especially with new control materials, and evaluating predictive models that estimate turfgrass insect activity periods. He has presented numerous seminars and talks to the turfgrass industry on turfgrass insects. His work has been published in scientific journals and trade publications. He is a coauthor of three books, *Managing Turfgrass Pests*, *Turfgrass Insect and Mite Manual* and *Handbook of Turfgrass Insect Pests*.

Acknowledgments

This second edition took three years to complete and was truly a joint effort (including organization and layout) of the "Harry and Dave Team." Though most of the photos were taken by us, some are the efforts of Glenn Berkey, retired photographer of the Ohio Agricultural Research and Development Center, The Ohio State University, Wooster, Ohio. Others were contributed by fellow entomologists and our friends in the Turfgrass Industry. Each of their photos are acknowledged in the book.

We appreciate the comments of our reviewers: Paul Latshaw, CGCS; Tom Walker, CGCS; "Jerry" Faubel, CGCS; Doug Halterman, Senior Vice-President, Leisure Lawn, Inc.; Dr. Garry Seitz, Leisure Lawn, Inc.; Dr. Bobby Joyner, Director of Research, Tru-Green/Chemlawn Tech Center; Michael Kernegham, Vice-President, Turfgrass Management Systems - Weed Man - Canada; Barbara Bloetscher, The Ohio State University; Bruce Shank, BioCom Horticultural Communications; Jean Steva, The Ohio State University; and Renee Shetlar.

The support of The Ohio State University, the Turfgrass Industry and the Ohio Turfgrass Foundation whose funding and encouragement helped facilitate much of the research that provided the experience upon which these writings are based is gratefully acknowledged.

This book is dedicated to our wives, Dolores A. Niemczyk and Renee Shetlar, who put up with us during the course of this effort.

JMJ. Thanks be to God.

Harry D. Niemczyk, Ph.D.
David J. Shetlar, Ph.D.

Contents

Page   Page

Contents (continued)

Notes

Introduction

We offer this practical book to turfgrass owners, managers, lawn service operators, golf course superintendents, students and others involved with turfgrass management with the hope of providing accurate identification of turfgrass pests and offering approaches for their control. This book is a compilation of color photos, illustrations and information, some of which we have published in various journals and books and presented in seminars. **Much is also based on our 50+ years of collective experience as turfgrass entomologists**. Other information was obtained from the publications of other entomologists.

Using This Book

Organization. The book is organized to lead the user to think about **turfgrass insects in relation to the specific segment of that environment which they occupy**; namely, LEAVES and STEMS, CROWNS, THATCH, or SOIL. This way of thinking **focuses on where the insect lives** and provides a logical basis for selecting the best approach when control measures are necessary. The book is written in easy-to-read language with limited text and avoidance of scientific terms and names. **Emphasis** is placed on the use of high quality color photos and line drawings to meet our objective of increasing the reader's knowledge of and ability to identify the insects and insect damage symptoms that occur in the turfgrass environment.

Pest Identification. The following steps are suggested when using this book to identify an insect, mite or other arthropod that you might have found in turfgrass.

1. Determine in which of the four environmental segments it was found: LEAVES and STEMS, CROWNS, THATCH, or SOIL.

2. Turn to the pictorial guide in Chapter 2 and find the color photo of the pest and/or symptom that matches what you see.

3. After identifying the insect or symptoms, refer to the page reference in the chapters that discuss the pest or symptom in greater detail.

Specimen Samples. If the insect found does not appear in this book, send several sample specimens to the state Cooperative Extension entomologist who is usually located at the state Land-Grant University, or to other professional entomologists with knowledge of turfgrass insects. Further details are given in **Chapter 9**.

Life Cycles. In **Chapters 3, 5 and 6**, color photos, identifying characteristics, life cycles and description of damage symptoms for diagnosis are given for insects in the SOIL, THATCH, CROWNS, LEAVES and STEMS, respectively. Refer to these life cycles for ideas about times that the pests are susceptible to control. These chapters also contain more information on pest descriptions and characteristic damage symptoms.

Control. Except where especially relevant, we have avoided listing specific products in this book because they frequently change, as do regulations about their use. Instead, **Chapters 4 and 7** are primarily devoted to the **principles** of using chemicals, biological agents and cultural methods to control the pests in their respective segments of the turfgrass environment. Development of a Pest Spectrum and Target Calendar, integration of control approaches and specific programs are outlined in **Chapter 8**.

The reader is reminded that while insecticides and other approaches to control change with new research findings, the PRINCIPLES underlying achievement of control remain unchanged.

When a new insecticide is labeled and becomes available, as much as possible should be learned about the product, including its **residual properties**, **mobility** in the turf and thatch environment, and **range of effectiveness** against the spectrum of pests that inhabit the soil, thatch, crowns, leaves and stems. This information, together with a knowledge of the pest's life history in the region and application of the **TARGET PRINCIPLE**, provides a sound basis for deciding if and when the product might be effective. Updated information on insecticides and other means of control is available from the Cooperative Extension Service in each state.

Detection and Monitoring. *Chapter 9* is devoted

to equipment and methods useful in detecting and monitoring infestations of turfgrass insects. Monitoring and detection are essential to any program to control pests. Therefore, this Chapter should be particularly useful. The methods and equipment illustrated have been used successfully by entomologists and turfgrass managers for many years.

Pest Damage Symptom Similarity. *Chapter*

10 presents a series of photos and descriptions of symptoms commonly observed in turfgrasses. While this series is a review of previously identified pests and associated symptoms, it is primarily intended to **emphasize** that symptoms caused by drought, improper or inadequate fertilization, diseases, winter desiccation, ball marks and a number of other factors can be similar to those caused by insects and mites. **Proper diagnosis is the point.**

Appendices. *Chapter 11* contains a list of common and

scientific names for the insects and other organisms covered in this book. Lists of useful reference books, trade magazines and other publications where specific information on turf insects can be found are included. Sources are also given for slide sets along with addresses of supply firms where detecting and monitoring equipment can be obtained. Also included is a listing of insecticides that have been, are currently, or are likely to be registered for turf usage.

We sincerely hope you will find this book useful. **Your comments, corrections, and suggestions for improving future editions are invited.**

Authors' Perspectives

Integrated Plant Management

Integrated Plant Management can be defined as a system of growing plants that utilizes and integrates a broad range of strategies, methods, and materials to establish and maintain healthy and vigorous plants. In *Turfgrass Management*, the focus is on the turfgrass **plant**, and the objective is to produce a stand of plants that is aesthetically and functionally acceptable to people. Insects and mites, and other pests, such as weeds and disease organisms, can influence the quality and therefore acceptability of the turfgrass site. However, insects are but one factor among the many components of the total *Plant Management System* that need to be integrated in order to produce the desired result. In other words, our perspective is that the **concept is Integrated *Plant* Management** and *Insect Pest Management* is only one component of that concept.

Insect Pest Management

The objective of Insect Pest Management is to utilize options, such as prevention, monitoring, trapping, natural predators, parasites, biological agents, and insecticides, to **control** insect pest populations so as to avoid or minimize damage that reduces turfgrass quality. Therefore, in this book, **control** rather than management is most often used to describe strategies or approaches for dealing with destructive insect and mite pests of turf.

Approaches to Control

One approach to insect control is to focus on suppression of developing or anticipated pest populations to prevent damage. This approach uses monitoring (observations) and control strategies including: application of naturally occurring and artificially produced biological agents or insecticides; distribution of natural predators and/or parasitic organisms; use of resistant - tolerant varieties; other cultural methods and materials; and, habitat modification and quarantine to prevent insects from damaging turf. This approach is referred to as the "***Preventive Approach.***"

The second approach, sometimes referred to as the "**Curative Approach**," requires the turf user/owner or manager to observe, survey, monitor, and map the turf for the purposes of locating, identifying and assessing insect populations and/or symptoms of injury. Only when further population development and/or unacceptable damage is anticipated, should control measures be employed. The objective is to prevent, stop or, at least, limit damage.

Actually, the objective of **both** approaches is to prevent (avoid) or, at least, limit (minimize) damage.

A third approach might be "**Tolerance.**" In this approach, Plant Management levels might vary, but unless damage is aesthetically unacceptable to the user/owner or interferes with the purpose of the turf site, it is simply tolerated.

Factors Affecting Control Approaches

1. Turf Quality Standards. The standards of turf quality vary with the quality requirements and/or expectations of the end-user or owner, standards imposed by those requiring the service of a turfgrass manager, and the personal standards of the turfgrass manager or service provider. All these factors influence the approach taken.

2. Financial Considerations. The level of financial resources committed to insect control is usually a major factor in the approach taken. Budget allocations for this part of Turfgrass Management vary greatly among golf course superintendents, athletic field and park managers, home owners, and lawn service operators who must realize a profit from services they perform.

Pest species occur and reoccur (e.g., cutworms on golf greens) at different times during the growing season. A **curative approach** usually targets **one pest at a time** and thus may require more than one treatment. The **preventive approach** usually targets control of **more than one pest** insect, thus potentially reducing the number of applications and overall cost of the program.

3. Pest Spectrum. Generally, the spectrum of insect pests of the area or site has major influence on the approach used. At sites with a history of soil-inhabiting pests, such as grubs or mole crickets, the preventive approach is generally followed. Areas where crown-thatch and leaf-stem inhabiting insects are primary pests, the curative approach is used most.

4. Manager's Perspective. Background, training, experience, consumer attitudes and public opinion influence individual philosophies and attitudes among turfgrass managers, owners and service providers and, therefore, the approach to control they follow.

Multiple Target Concept

The first edition of **Destructive Turf Insects** and this second edition provide information on individual insects and mites according to the segment of the turfgrass environment they occupy. However, the reality of experience commonly demonstrates that **INSECT AND MITE PESTS OF TURFGRASS RARELY OCCUR "ONE AT A TIME AT ANY ONE TIME."** Development of cultural programs, materials, methods and timing of treatments should include consideration of the **complete spectrum of pests** occurring at the time of application. While the focus of an application may be on a specific target insect(s), the treatment can also impact the damaging stage of another insect that occurs later in the growing season.

Consideration of the impact on the total pest complex any time a treatment for control is made is to apply the "**Multiple Target Concept**."

Application of the Multiple Target Concept requires full knowledge of the spectrum of pests affected by the control agent applied. **A good start** is to list all potential target pests and their stages each month of the growing season, then consider the impact of the treatment on present and future stages of each pest if applied during each of the months.

In view of the development of new classes of control agents with relatively long residual effectiveness against soil-inhabiting pests, their affect on the spectrum of pests should be known. This consideration can lead to a **reduction in the number of applications** necessary to prevent damage.

Chapter 2

Pictorial Guide to Some Destructive Insects, Mites & Other Pests of Turfgrass

Soil Inhabitants

Grub Damage

Patches of wilted, dead, or dying grass; sod loose from soil, often damaged by skunks, raccoons, armadillos or birds searching for grubs.

Grubs   pp. 17-31, 41, 89

Grubs

Larvae white; brown to tan head, C-shaped; with small legs; found feeding on roots at thatch-soil line.

Grubs   pp. 17-31, 41, 89

Black Turfgrass Ataenius

Small black beetles, 3/8 inch long; some red-brown before maturing; wide front legs; larvae small, C-shaped, with small legs; feed on turfgrass; primarily a pest of golf courses.

Black Turfgrass Ataenius
                  pp. 17-19, 27-29, 41, 89

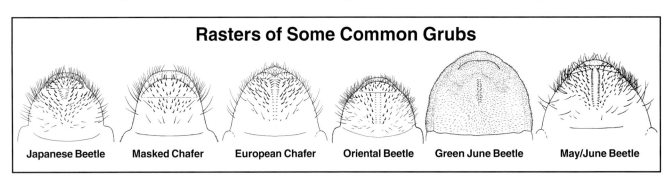

Rasters of Some Common Grubs

Japanese Beetle   **Masked Chafer**   **European Chafer**   **Oriental Beetle**   **Green June Beetle**   **May/June Beetle**

Green June Beetle Damage

Mounds of soil appearing in early spring or late fall; soil loose around turf roots but turf not easily lifted; large diameter holes extending into soil.

Green June Beetle
                  pp. 17-19, 25-27, 41, 89

Green June Beetle Grub

Large white grub found on surface, especially after rain; <u>crawls on back</u>.

Green June Beetle
                  pp. 17-19, 25-27, 41, 89

Earthworm Castings

Small to medium mounds of soil, usually appearing as a mounded cord; re-occurs regularly after rain or irrigation.

Earthworm Castings

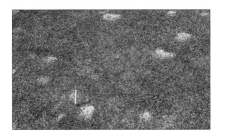

Ant Mounds

Small to medium, usually volcano-shaped, mounds of loose soil.

Ant Mounds   pp. 36-37, 41, 89

Fire Ants

Medium to large, rounded top, mounds of soil appearing in southern turf.

Fire Ants   pp. 36, 37-38, 41, 89

Cicada Killer Wasp Burrow

Medium to large, irregular mound of soil in turf, usually with a channel leading to a hole. <u>Large wasp often flying nearby.</u>

Cicada Killer   pp. 38-39, 41, 89

Mole Cricket Damage

Thinning turf with underlying soil loose; elongate trails and emergence holes commonly seen.

Mole Cricket   pp. 31-35, 41, 89

Mole Crickets

Gray-brown winged insects with short, spade-like front legs; forcefully push through fingers when held; mainly southern turf pest.

Mole Cricket   pp. 31-35, 41, 89

Ground Pearls

Round yellow to purple spheres ranging in size from a grain of sand to 3/16 inch in diameter; grass yellow to brown in irregular patches; a southern pest of centipedegrass.

Ground Pearls   pp. 35-36, 41

Identification of Mole Crickets by Front Legs

Inspect "toes" (=dactyls) of front leg.

Four "toes"
Native (=Northern)

Close together "toes"
Tawny

Widely spaced "toes"
Southern & Shortwinged

Crane Flies - Leather Jackets

Gray to brownish-white maggots, head capsule indistinct, no legs, and <u>finger-like projections</u> surrounding tip of abdomen. Causes thinning turf in irregular patches

Leather Jackets/Crane Fly   pp. 39

Moles

Tunnels under turf, one inch diameter or larger, raised into ridges; occasional large mounds of soil pushed up through turf.

Moles

Voles and Shrews

Tunnels in turf and thatch; often appearing after snow cover melts; may extend into flower beds or under mulch cover.

Voles or Shrews

11

Crown and Thatch Inhabitants

Sod Webworm Adult

Gray-tan moths with <u>snout-like projections</u>; wings rolled around body; fly over turf at dusk; fly erratically for short distances when disturbed.

Sod Webworm Adult

pp. 53-58, 83, 89

Spring Sod Webworm Damage

Small to large areas that do not green up in spring; often probed by starlings.

Sod Webworm

pp. 53-58, 83, 89

Sod Webworm on Green

Crescent shaped, brown marks appear on golf green or tee; consist of top dressing held together with silk over a burrow.

Sod Webworm

pp. 53-58, 83, 89

Sod Webworm

Greenish to beige-brown to gray larvae, most with circular to rectangular spots; live in silk-lined tunnels; grass chewed off close to crown; birds frequent infested areas.

Sod Webworm

pp. 53-58, 83, 89

Armyworm

Brown to gray-green larvae clearly marked with light stripes; brown head with darker brown markings; hide in thatch during day and feed at night, cutting off grass close to crown; often in groups.

Armyworm

pp. 64-65, 66, 83, 89

Fall Armyworm

Green to brown larvae clearly marked with light and dark stripes; <u>head with inverted, white, Y-shaped mark</u>; feed at night, cutting off grass close to crown; often in groups.

Fall Armyworm

pp. 64-65, 66-67, 83, 89

Black Cutworm

Dark olive to gray, thick-bodied larva with one pale stripe on upper surface; common pest of golf greens; hide in holes or thatch during the day.

Black Cutworm

pp. 64-65, 83, 89

Black Cutworm Damage

Circular or finger shaped sunken areas on golf course greens or tees; turf eaten below mowing level; often around aerification holes.

Black Cutworm

pp. 64-65, 83, 89

Bronzed Cutworm

Thick-bodied, brown larva with distinct bronze sheen; three white stripes on upper surface, especially on dark collar behind head; broad, pale white to yellow stripe on each side; feed in early spring.

Bronzed Cutworm pp. 64-65, 68, 83

Hairy Chinch Bug
Small, black bugs, 1/5-inch long with white wings folded over back; some with short wings; nymphs (young) red with white band across body.

Chinch Bugs   pp. 69-70, 83, 89

Chinch Bug Damage
Irregular spots of yellowish turf; grass appears droughty; grass may be killed in large or small areas.

Chinch Bugs   pp. 69-71, 83, 89

Bigeyed Bug
Gray to black bug, 1/5-inch long with blunt head and protruding eyes; move rapidly; wider than chinch bugs, but often found with their infestations.

Bigeyed Bug   pp. 71

Billbug Larva
Small, white legless larva with brown head, feeding inside stems, at the crown or on roots; fine white sawdust-like material often present.

Billbugs
                pp. 59-62, 72, 83, 89

Bluegrass Billbug Damage
Numerous small patches of dead grass; often confused with drought and sod webworm damage; plants easily broken off at crown; a pest of Kentucky bluegrass and perennial ryegrass.

Bluegrass Billbug
                pp. 59-61, 83, 89

Billbug Damage
Small to large patches of dead or dying grass; evidence of tunneling in stems that easily break off at the crown.

Billbugs
                pp. 59-62, 72, 83, 89

Annual Bluegrass Weevil
Small mottled to shiny brown-black snout beetle, 5/32-inch long; chew notches in edges of blades and holes in center of stems of annual bluegrass; New England, NY, PA, NJ & WV.

Annual Bluegrass Weevil
                pp. 63-64, 83, 89

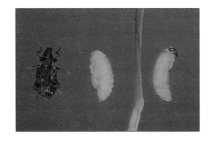

Annual Bluegrass Weevil
Only on annual bluegrass; small, creamy-white, legless larva with brown head; feed at base of plants, chewing out U-shaped notches.

Annual Bluegrass Weevil
                pp. 63-64, 83, 89

Twolined Spittlebug
Frothy, spittle-like masses in thatch, at base of grass stems; soft, tan to green insect inside; occurs in centipedegrass and bermudagrass in spring.

Twolined Spittlebug   pp. 81, 83, 89

Leaf and Stem Inhabitants

Greenbug
Small soft-bodied insects 1/16-inch long; light green body with darker stripe down back; ten to 50 winged and/or wingless forms occur on a single grass blade.

Greenbug   pp. 80, 83, 89

Greenbug Damage
Grass showing yellow to burnt orange coloration, often under trees but also in open areas; may resemble leaf rust disease.

Greenbug   pp. 80, 83, 89

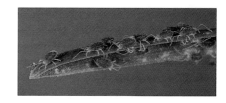

Clover Mite
White to straw colored damaged grass, often next to house foundations or near tree shade lines; tiny pinkish to brown-bodied mites with eight legs in turf, front pair of legs extend forward from body.

Clover Mite   pp. 77, 83, 89

Clover Mite Damage
Grass appears frosted and desiccated in fall or spring, often near buildings where turf is grown up to walls.

Clover Mite   pp. 77, 83, 89

Winter Grain Mite
Grass appears desiccated in late fall or early spring; mites active in winter, under cover of snow; olive to black body with eight bright red legs; anus on upper surface of body.

Winter Grain Mite
                              pp. 77-78, 83, 89

Banks Grass Mite
Straw colored damaged grass surrounding trees in spring in Rocky Mountain States; general yellowing of grass in summer in southern states.

Banks Grass Mite   pp. 76, 83, 89

Bermudagrass Mite
Bermudagrass shows tufts or rosettes at internodes; microscopic elongate, creamy-white mites in leaf sheaths.

Bermudagrass Mite
                              pp. 75-76, 83, 89

Rhodesgrass Mealybug
White, cottony structures, 1/8-inch in diameter at bases of grass stems and in leaf axils; southern grasses.

Rhodesgrass Mealybug
                              pp. 79, 83, 89

Bermudagrass Scale
White, oval structures, 1/15-inch long clustering at nodes of bermudagrass or attached to rhizomes; turf turns brown in summer and delays greenup in spring.

Bermudagrass Scale
                              pp. 78-79, 83, 89

Frit Fly

Small, shiny, black flies 1/16-inch long with yellow on legs; attracted to white surfaces such as golf balls; tiny larvae tunnel in grass stems causing death.

Frit Fly   p. 81-82, 83, 89

Billbug Larva

Small, white legless larva with brown head, feeding inside stems, at the crown or on roots; fine white sawdust-like material often present.

Billbugs   pp. 59-62, 72, 83, 89

Billbug Damage

Small to large patches of dead or dying grass; evidence of tunneling in stems that break off easily at the crown.

Bluegrass Billbug
                              pp. 59-61, 83, 89

Spider Webs

Small, usually circular webs on turf surface; evident in the morning when covered with dew; mistaken for sod webworms which do form patches of webbing on surface.

Spider Webs

Annual Bluegrass Weevil

Small mottled to shiny brown-black snout beetle, 5/32-inch long; chew notches in edges of blades and holes in center of stems of annual bluegrass; New England, NY, PA, NJ & WV.
Annual Bluegrass Weevil
                              pp. 63-64, 83, 89

Annual Bluegrass Weevil

Only on annual bluegrass; small, creamy-white, legless larva with brown head; feed at base of plants, causing annual bluegrass to brown in June and July.
Annual Bluegrass Weevil
                              pp. 63-64, 83, 89

Twolined Spittlebug Adult

Boat-shaped, black insects with <u>two red-orange stripes on the wings</u> jump and fly from turf when disturbed.

Twolined Spittlebug
                              pp. 81, 83, 89

Leafhoppers

Wedge-shaped insects, green to brown, jump and fly short distances in turf; often in great numbers, but not a serious pest.

Leafhoppers

Sod Webworm Adult

Gray-tan moths with <u>snout-like projections</u>; wings rolled around body; fly over turf at dusk; fly erratically when disturbed by mowing or walking across turf.
Sod Webworm Adult
                              pp. 53-58, 83, 89

Notes

Chapter 3

Soil-Inhabiting Pests

Grubs in General

Grubs feed in soil, thatch and at the thatch-soil interface where they consume partially decomposed thatch, other organic matter, soil particles and living turfgrass roots.

White grubs are the larvae of many species of beetles, mainly belonging to one family - the scarabs. Adults differ in color markings, habits and life cycles, but grubs are similar in appearance. Fully grown larvae are 1/2- to 3/4-inch long, white to grayish, with brown

heads and six distinct legs. They usually assume a C-shaped position in the soil. Severe infestations feeding in the soil-thatch interface of turf can destroy most of the roots, causing the turf to turn brown and die. Moles, birds, raccoons, skunks and armadillos actively feed on grubs, and in the process, tear up the turf as they search for them.

Life Cycles. The life cycles of grubs can be classified according to the time required for completion of the cycle from egg to adult, namely:

1. Less than one year
2. One year
3. Two years or more

Among those requiring two to five years to complete a cycle are some species of May beetles or "Junebugs," often seen around lights on warm spring nights. These beetles generally are large, hard-bodied, and vary in color from tan to brown to black. Some species feed on the foliage of trees and shrubs at night, others do not feed at all. Feeding, flight, mating and egg laying occur at night.

Grub pests of turf (left to right): May/June beetle, green June beetle, European chafer, masked chafer, Japanese beetle, Oriental beetle, Asiatic garden beetle, and black turfgrass ataenius.

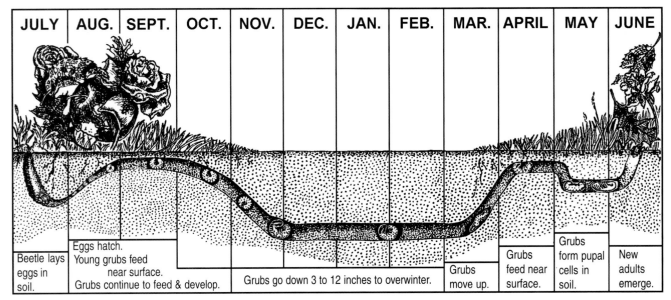

JULY	AUG.	SEPT.	OCT.	NOV.	DEC.	JAN.	FEB.	MAR.	APRIL	MAY	JUNE

| Beetle lays eggs in soil. | Eggs hatch. Young grubs feed near surface. Grubs continue to feed & develop. | | Grubs go down 3 to 12 inches to overwinter. | | | | | Grubs move up. | Grubs feed near surface. | Grubs form pupal cells in soil. | New adults emerge. |

Seasonal life cycle of the Japanese beetle, a typical one-year life cycle.

Masked chafers, Oriental beetle, Asiatic garden beetle, European chafer and green June beetle are examples of grub species that generally complete their life cycle in one year. The common Japanese beetle is perhaps the best known example of this group.

The black turfgrass ataenius grub often has two generations per year, especially in the southern part of its range.

White grub damage to turf (pulled back to see grubs).

Grub Diagnosis.
Evidence of annual white grub damage includes patches of wilted, dead or dying turf visible during spring (April and May) and fall (September to November). The black turfgrass ataenius may cause similar symptoms in late June, July or August. During these periods, the presence of the grubs is often made evident by the feeding activity of skunks and other mammals which tear up the turf in search of grubs. Large flocks of various black birds (e.g., crows, starlings, grackles) often feed on grubs in heavily infested areas.

Ground mole activity may also indicate white grub problems, though moles also feed on earthworms or other insects.

The sure way to detect white grubs is to cut into the turf in four or five locations and examine the root zone and first three inches of soil carefully. A standard golf course cup cutter can be used for this purpose. The sample can be examined and replaced in the hole without complete destruction. Pouring water on the replaced sample helps survival of the disturbed turf.

Methods and equipment for mapping and surveying for grubs are given in Chapter 9.

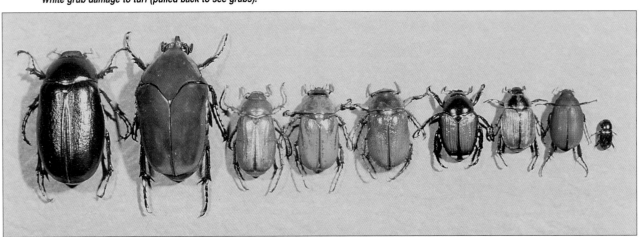

Adult stage of turf grub pests (left to right): May/June beetle, green June beetle, European chafer, southern masked chafer, northern masked chafer, Japanese beetle, Oriental beetle, Asiatic garden beetle, and black turfgrass ataenius.

Grub Identification.

There are many species of white grubs, and identification is based primarily on the pattern of spines found on the <u>underside of the tip of the abdomen</u>. This area is called the **raster**. A 10 to 15-power hand lens is adequate for examining the rastral pattern of most white grubs. If the grubs are very small, a dissecting microscope with 20 to 40-power magnification may be needed.

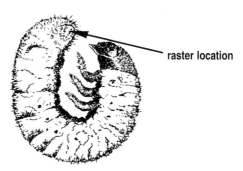

raster location

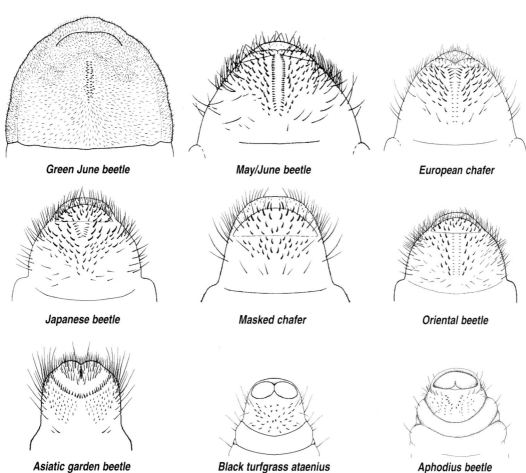

Green June beetle

May/June beetle

European chafer

Japanese beetle

Masked chafer

Oriental beetle

Asiatic garden beetle

Black turfgrass ataenius

Aphodius beetle

Raster Patterns of Common White Grubs

Skunks dig in turf and feed on grubs.

Moles commonly search for and feed on grubs. They feed primarily on earthworms, so their activity often, <u>but not always</u>, indicates grubs are present.

Japanese Beetle

The Japanese beetle is an imported pest generally found east of a line roughly running from Michigan, southern Wisconsin and Illinois south to Alabama. Established, localized populations are also known in Nebraska, Iowa, Missouri and northern Arkansas. Occasional introductions are made into western states such as California, Nevada and Oregon when the adult beetles or larvae are shipped in commerce. The original population was detected in New Jersey in 1916 having been introduced from Japan.

Japanese beetle adults skeletonize leaves of many plants by eating the soft tissue and leaving the veins. Blooming roses and buds are favorite food plants.

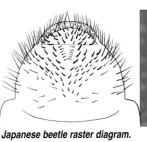

Japanese beetle raster diagram.

Photo of Japanese beetle raster.

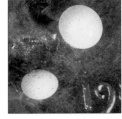

All grub adults lay dehydrated eggs (bottom) that must absorb moisture from surrounding soil in order to develop (top). Droughty turf/soil at egg laying usually has fewer grubs because of poor egg survival. Thus, withholding irrigation at egg laying may help suppress grub infestation.

The adult beetles are general plant feeders known to feed on over 400 species of broad-leaved plants, though only about 50 species are preferred. The grubs will also feed on a wide variety of plant roots including ornamental trees and shrubs, and turfgrasses.

Adults are a brilliant metallic green beetle, generally oval in outline, 3/8-inch (8-11mm) long and 1/4-inch (5-7mm) wide and are strong daytime flyers. The wing covers are a coppery-brown color and the abdomen has a unique row of five tufts of white hairs on each side.

Diagnosis.
Turf infested with grubs first appears off color as if under water stress. Irrigation often masks infestations, but when withdrawn, damage soon appears. The turf feels spongy under foot and can be easily pulled back to reveal the grubs. Large populations of grubs kill the turf in irregular patches. Infested sites are often torn up by mammals and birds feeding on the grubs.

The larvae are typical white grubs which can be separated from other soil dwelling white grubs by the presence of a **V-shaped series of bristles** on the raster. First instar larvae are about 1/16-inch (1.5mm) long while the mature third instars are about 1-1/4-inch (32mm) long.

Life Cycle and Habits.
Adult beetles emerge during the last week of June through July. The first beetles out of the ground seek out suitable food plants and begin to feed as soon as possible. These early arrivals release an aggregation pheromone (odor) which attracts additional adults to gather in masses on the plants first selected.

Newly emerged females release an additional sex pheromone that attracts males. The first mating usually takes place on the turf with several male suitors trying to mate with one female. "Balls" of beetles are often seen on the turf surface in the early afternoon. Mating is also common on food plants and several matings by both males and females are common. After feeding, the females leave feeding sites in the afternoon and burrow into the soil to lay 1 to 5 eggs at a depth of 2 to 4 inches. These females return to feed, mate and lay eggs until 40 to 60 eggs are laid. Most eggs (75%) are generally laid by mid-August though adults may be found until the first frost of fall.

If the soil is sufficiently moist, the eggs will swell in a day or two and egg development takes 8 to 9 days at 80 to 90°F or as long as 30 days at 65°F. The first instar larvae dig through the soil and thatch where they feed on roots, thatch, soil, and organic material.

During development, grubs may tunnel laterally, feeding on organic matter and fresh roots, creating a spongy feel to the soil and turf. Most of the grubs are in the third and final instar by mid-September and, as soil temperature drops by mid-October, burrow into the soil to spend the winter. Grubs can survive being in soils that do not get below 25°F. Grubs have been recovered in Canada during the winter months at a depth of 30 inches!

The grubs return to the surface in spring as the surface soil temperature warms to 60°F, usually in mid- to late April. The grubs continue their development in spring and move down to pupate in late May and early June. Adults emerge from late June through mid-July.

A single female Japanese beetle releases a pheromone attracting many males who attempt to mate with her, creating a ball of beetles.

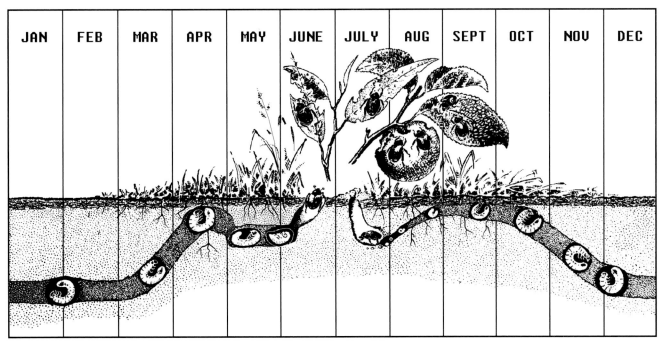

| JAN | FEB | MAR | APR | MAY | JUNE | JULY | AUG | SEPT | OCT | NOV | DEC |

Japanese Beetle Life Cycle in Ohio.

Masked Chafers

Masked chafers are natives of North America. The southern masked chafer is most common on bermudagrass in the southern states and commonly attacks turfgrass in transition zones (Kentucky bluegrass and tall fescue), especially where Japanese beetle grubs have been suppressed. It has been recently reported as a more common pest in southern Indiana, Illinois, Iowa and Nebraska. The northern masked chafer is commonly a pest in the Kentucky bluegrass and perennial ryegrass growing regions from New England across to Illinois. Southwestern masked chafer is a common pest in Texas and Oklahoma to southern California. The western masked chafer is common in northwestern Oklahoma and western Kansas into northern California.

Adults do not feed.

Adults are a dull dark yellow-ocher and have a darker brownish-black band on the head that joins the eyes together to form a black mask. Adults of the northern masked chafer have conspicuous hair on the thorax and wing covers. The southern masked chafer has sparse hair. Adults are 1/2-inch (11 to 14 mm) long by 1/4-inch (6 to 7 mm) wide.

Diagnosis.
Turf begins to show what appears to be drought stress in late summer into fall or spring and does not recover after rain or irrigation. Heavy infestations result in turf dying in irregular, large or small, patches. Birds, skunks, raccoons and opossums commonly dig up turf around the dead patches. Moles may tunnel extensively where grub populations are high. Infested turf feels spongy under foot and is easily lifted because of the absence of roots. The adults do not feed on ornamental plants or turf.

First instar larvae are about 3/16-inch (4.5 mm) long at hatching and reach one inch (22 to 25 mm) when mature third instars. The

mouthparts must be dissected to distinguish between the northern and southern masked chafers. The larvae have an irregular pattern of bristles on the raster, typical of all masked chafer larvae.

Masked chafers emerge and mate at night. Virgin females release a pheromone that attracts males.

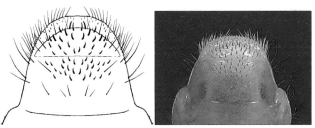

Masked chafer raster has no pattern. Photo of masked chafer raster.

Life Cycle and Habits. Northern and southern masked chafers have very similar life histories and habits. Adult beetles usually begin emergence in mid-June and are active into mid-July. Males come to the soil surface after dark before females emerge. The southern masked chafer is apparently active earlier in the evening than the northern masked chafer. Southern masked chafer males begin to emerge just before sunset and fly near the ground surface in search of unmated females. Most mating and flying activity is finished by midnight.

Northern masked chafer maximum activity occurs around midnight. Unmated females come to the soil surface, climb upon a grass blade and begin releasing a sex pheromone to attract males. Many males often cluster around calling females during copulation. Mated females and males fly at night and are strongly attracted to lights. Males tend to fly within two feet of the ground while females seem to fly higher. Northern masked chafers are active until a few hours before sunrise

Neither males nor females feed on plant material. Mated females dig down 4 to 6 inches and lay 11 to 14 eggs. If soil moisture is sufficient, the eggs swell within eight days and hatch in 14 to 18 days at 70 to 75°F. Young larvae burrow to the soil surface in search of plant roots, organic material as well as thatch. The second instar is reached in 20 to 24 days at 80°F and third instars are common by late August when most of the damage occurs.

As the soil temperatures drop in the fall, the larvae burrow downward, some to 12 inches, but in southern states most are within 3 to 6 inches of the surface. Grubs surviving the winter return to the upper level of soil in late April and May to feed and again move down in late May and early June to pupate. The pupa takes about 17 days to mature and become an adult.

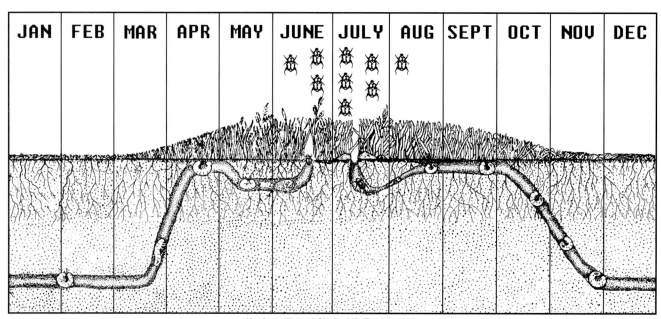

| JAN | FEB | MAR | APR | MAY | JUNE | JULY | AUG | SEPT | OCT | NOV | DEC |

Northern Masked Chafer Life Cycle in Ohio.

European Chafer

The European chafer was imported to the United States and first detected in Newark, NY in 1940. Since then the pest has spread into Connecticut and upper New York, west to Ohio and south from West Virginia across Maryland. A population is well established in Michigan and southern Ontario, Canada.

European chafer adults appear similar to some species of May/June beetles. Examination of the tarsal claws properly identify each type of beetle.

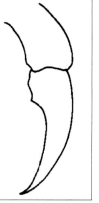

European chafer tarsal claw with no tooth.

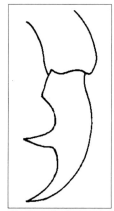

May/June beetle tarsal claw with distinct tooth.

Adults look much like some of the light colored June beetles. However, the European chafer is 1/2-inch (13 to 14 mm) long, shorter than most June beetles, and the wing covers have distinct longitudinal grooves. The most distinctive characteristic is the absence of a tooth on the tarsal claw of the middle leg. May or June beetles, have a distinct tooth. The adults do not feed.

The grubs feed on organic matter and a wide variety of plant roots, such as clover, small grains, and weeds. They are major pests of all cool-season turfgrasses and nursery stock.

Diagnosis. Typical grub damage of thin turf, wilting and death in irregular patches can be found in the fall and early spring.

Life stages are typical of annual white grubs and generally there is one generation per year. From 0.5 to 1% may require two years to complete a generation.

First instar larvae are about 3/16-inch (4 mm) long and are approximately 1-1/8-inch (27 mm) long when in the third instar. All three instars are typical C-shaped white grubs but these can be identified by the raster that has two parallel rows of bristles that **diverge laterally below the anus**. When observed "end on," the anal opening is Y-shaped. These grubs are slightly smaller than the May or June beetle grubs which also have two parallel rows of bristles on the raster that **do not diverge** at the anus.

Life Cycle and Habits. Adults begin emerging from pupal cells in mid-June and continue mating and ovipositing until late July. Most activity occurs from the last week of June through the second week of July. Adults emerge at sundown and fly to nearby trees and shrubs silhouetted against the sky. Here, large numbers fly, with a considerable buzzing noise, for 20 to 35 minutes. Thousands may congregate on a single tree. When the sky is truly dark, adults settle on the foliage and begin copulation. Mating continues in mass until daybreak when the adults burrow into the soil. Cool or rainy nights greatly reduce flight and mating activities. Apparently adults may return to trees several times for mating, but eventually females dig into the soil to lay eggs.

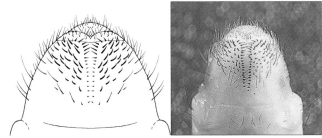

Diagram of European chafer raster.   *Photo of European chafer raster.*

Each female lays 15 to 20 eggs in 2 to 5 days. The eggs are usually laid singly in compacted cells of soil 2 to 6 inches deep. The eggs swell as they absorb soil moisture and hatch in about two weeks. First instar larvae may remain deep in the soil if surface soil moisture is low. Eventually, the young larvae move to the surface and feed on organic matter, including thatch and plant roots. If food is sufficient, the first instar matures in about three weeks. The second instar takes about four more weeks to mature. This pest may move up, down and laterally depending upon soil moisture and food availability. The third instar feeds for a period in the fall before moving down for the winter.

European chafer grubs are known to feed longer in the fall than many other species before moving down and are one of the first to return to the surface in the spring, often in March. In thick sod and under heavy snow cover, grubs may remain in the upper one to two inches of soil during the winter. Pupation occurs in mid-May, 2 to 6 inches in the soil.

While this species generally has one generation each year, from 0.5 to 1% of the grubs has been known to require two years to develop into adults.

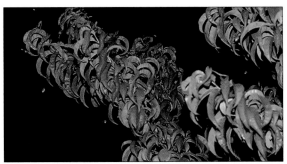

Adult European chafers fly to and land on trees at sundown creating a buzzing sound. The beetles mate on trees but do not feed on the foliage.

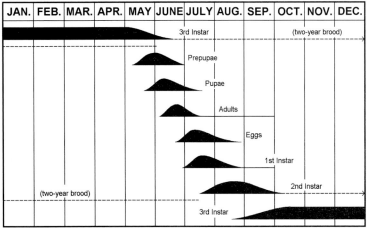

European Chafer Life Cycle in New York (redrawn from Tashiro et al.).

Oriental Beetle

This native of Japan was first detected in the United States in Connecticut in 1920 and has moved little from that time to the mid-1970s. It has since moved, mainly in soil of nursery stock, to periodically damage turf in northeastern Ohio, New York, Pennsylvania, New Jersey, most of the New England States and into the Carolinas.

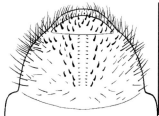

Diagram of Oriental beetle raster.   Photo of Oriental beetle raster.

Oriental beetle adults feed on daisies and other flowers. Note the three common color forms.

The larvae readily attack the roots of cool season turfgrasses, such as Kentucky bluegrass and fine fescues, and often occur in mixed populations with the Japanese beetle. Occasionally, they will feed on roots of weeds and are known to damage full grown nursery stock and outdoor potted plants.

Adults are 3/8- to 7/16-inch long (9 to 10 mm) and vary considerably in intensity of the markings on the thorax and wing covers. Individuals may be entirely brownish-black to entirely straw colored except for a brown head and mark on the pronotum. Usually the head is solid dark brown, the pronotum is dark in the center outlined in straw color, and the wing covers have longitudinal grooves and are mottled with patterns of dark brown on straw. The head and prothorax are usually iridescent bronze in bright light. The adults feed little and are occasionally found on the petals of flowers, especially daisies.

Diagnosis. Grubs feed on turf roots, thatch, and organic matter close to the surface (within one inch) causing typical grub damage consisting of wilting turf which may die in irregular patches. This pest prefers to feed on turf in sunny areas.

Mature third instar Oriental beetle larvae are approximately the same size, about 1-inch (25 mm) long, and externally resemble Japanese beetle grubs. The anal opening is transverse and the raster has two parallel rows of about 14 smaller bristles on each side. These two rows of smaller bristles usually have an adjacent row of longer, sometimes less distinctly lined up spines. This raster pattern can be confused with young May-June beetle larvae. However, the difference is that May-June beetle larvae have a broad V-shaped anal opening and that of the oriental beetle is a ***transverse curve***.

Life Cycle and Habits. The Oriental beetle has one generation per year. Adults begin to emerge in late June and some may be found as late as September. Active in July, adults may fly short distances in the morning but are most active in the evenings when they are commonly found chewing on petals of flowers. Some adults are attracted to lights but never in large numbers.

A few days after mating, the females burrow into the soil to lay an average of 26 eggs in small groups between 3 and 9 inches in the soil. The eggs must be deposited in moist soil so that water can be absorbed and development continued. At normal soil temperatures in July and August, the eggs take 18 to 24 days to hatch. The young grubs move to the soil surface to feed on roots and organic material. Second instar larvae are found in 3 to 4 weeks and usually molt into the third instar in 3 to 4 weeks. A majority of the larvae overwinter as third instars but some overwinter in the second instar. Larvae burrow down in late October and November and return to feed the following April and May. Pupae are present from mid- to late June and usually take about two weeks to mature.

Asiatic Garden Beetle

Asiatic garden beetle was introduced from Japan in the 1920s. It is most common in northeastern United States from New England to Ohio and into South Carolina.

Larvae occasionally attack turf but seem to prefer a variety of roots from weeds, flowers and vegetables. The adults feed on over 100 species of plants, preferring flowers of asters, dahlias, mums, roses and the leaves of a variety of trees and vegetables.

Adults are 5/16- to 3/8-inch (7 to 10 mm) long and broadly wedge shaped. The beetles are chestnut brown and often have a slight iridescent, velvety sheen. The abdomen protrudes slightly from

Asiatic garden beetles mate on tree leaves at night.

under the wing covers and the undersurface of the thorax has an irregular covering of short yellow hairs. The hind legs are distinctly larger and broader than the front or middle legs.

Diagnosis.

Larvae cause typical damage to turf, wilting and irregular patches of dead turf. Heavy infestations are uncommon but have been known to occur. Generally, **the deeper feeding habit** of Asiatic garden beetle grubs results in less damage to turf than from other species. With adequate rainfall or irrigation, populations of 20 per ft″ can be tolerated. Grubs prefer roots of other plants, therefore, they may be clustered around weedy areas especially near orange hawkweed. Grubs may also be concentrated next to flower beds where plants preferred by the adults are located.

Adults feed at night, notching and shredding foliage of trees, especially maples and elms, leaving plants in a ragged condition. They do not skeletonize leaves like Japanese beetles. Flowers, especially daisies, often have petals eaten off.

First instar larvae are about 1/16-inch (1.4 mm) long and have light brown head capsules. Full grown larvae are 3/8- to 1/2-inch (15 to 18 mm) long when extended. The grubs are commonly identified by the enlarged, light colored, maxillary palps (see photo below right) which appear to be in constant motion. AGB grubs can be identified by the **longitudinal anal slit and transverse curved row of brown spines** making up the raster.

Life Cycle and Habits.

Adults emerge at night and fly actively when temperatures are above 70°F from late June to the end of October, but most occur from mid-July to mid-August. Adults are strongly attracted to lights. During the day, beetles hide in the soil around favored food plants. The females tend to search out turf and pastures for egg laying and generally deposit an average of 60 eggs one to two inches deep in the soil. Eggs are laid over several weeks and normally hatch in 10 days during summer temperatures.

Young larvae dig to the soil surface where they feed on roots, thatch and organic material. Most first instar larvae are found in August and early September. Second instars are found in September and many do not reach third instar until the following spring. About half the population overwinter as second instars and the remainder as partially developed third instars. As cool October temperatures arrive, the larvae burrow down 8 to 17 inches to pass the winter.

Larvae return to the soil surface in spring and most mature by mid-June. They then pupate in compacted earthen cells, 1.5 to 4 inches in the soil. The pupal stage is relatively short, lasting 8 to 15 days. Adults remain in old pupal skins, changing from white to the mature chestnut brown a few days before digging to the surface.

Asiatic garden beetle mature grub. Note white palp base under head.

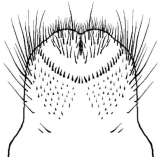
Diagram of Asiatic garden beetle raster.

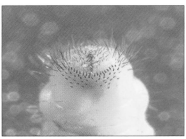
Photo of Asiatic garden beetle raster.

Green June Beetle

The green June beetle is native to North America, commonly found from southern Pennsylvania across to Oklahoma and south. It is most commonly a turf pest in the transition zones of Tennessee and Kentucky to the Carolinas.

The larvae feed on the roots of many species of turfgrasses and field crops and seems to prefer areas with high organic matter.

The adults are about 13/16- to 1-inch (20 to 25 mm) long, and generally are a velvety green color with orange-yellow margins. The lower surface is a shiny metallic green. The head has a distinctive flat horn.

Adults begin to emerge in late June but are most common in July and early to mid-August. Flight occurs during the daytime and the beetle makes a considerable buzzing noise that often alarms people. Adults may congregate around seeping wounds of trees, and are very common on ripe grapes, figs, peaches, plums, melons, some vegetables and ears of corn.

Green June beetle adults vary considerably in the amount of tan coloration on the wing covers. When flying, they expose a shiny iridescent green color.

Diagnosis. The <u>larvae</u> occasionally attack turf sufficiently to cause death in irregular patches. More commonly, larvae and adults make burrows in the turf, <u>pushing up mounds of soil somewhat resembling mounds made by ants</u>. Larvae feed on decaying organic matter and are known to consume green plant material. Damage to turf is caused primarily by their constant burrowing and tunneling rather than feeding on turf roots. The larvae also have the curious habit of **crawling about on their backs**! After rains, grubs are often seen crawling across sidewalks, driveways and are often found in swimming pools and garages.

The <u>larvae</u> are somewhat atypical of most turf grubs. They have the normal C-shape, but <u>usually move by stretching out on their backs to creep along with an undulating motion</u>. Because of this mode of movement, the legs are considerably smaller than other white grubs. <u>This movement is a specific characteristic of this species</u> (there is a related species that commonly inhabits mulch and compost piles which also crawls in this manner). First instars are 1/4-inch (6 to 6.5 mm) long, second instars 5/8-inch (15 to 17 mm), and mature third instars are 1-3/4-inch (45 to 48 mm).

Green June beetle larvae characteristically crawl across surfaces on their backs.

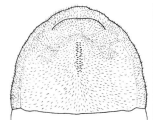

Diagram of green June beetle raster.

Photo of green June beetle raster.

Life Cycle and Habits. Green adults begin to emerge in late June, flying during the daytime, and are very common in July and August. The adults are attracted to ripe fruit and vegetables as well as seeping tree wounds. Once a feeding site is established, several adults may be actively flying about or attempting to mate. Newly emerged females call males to the ground for mating by producing a milky fluid that acts as an attractant.

Mated females, ready to lay eggs, dig a burrow pushing up a small mound of soil. Females excavate a small cavity in the soil 2 to 5 inches deep in which they lay 10 to 30 eggs. These **eggs are packed into a ball of soil** about the size of a walnut. Females may emerge and dig a new burrow elsewhere, but more commonly continue to dig 1 to 2 additional egg chambers in the first burrow. The eggs swell as they absorb moisture and hatch in 2 to 3 weeks.

Young grubs are very active and can move a considerable distance in a short period of time. Since eggs are clustered, infestations appear to be in groups or colonies. The grubs feed on organic matter and some green tissue, but their habit of constant burrowing and tunneling loosens the soil causing plant desiccation, the primary cause of damaged turf. Second instar larvae construct individual burrows which average 6 to 12 inches deep. At night, the

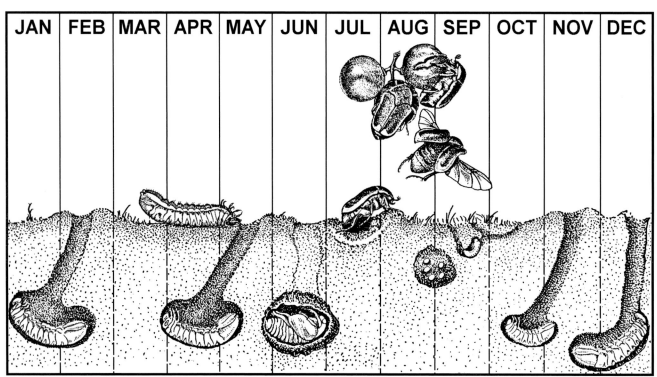

| JAN | FEB | MAR | APR | MAY | JUN | JUL | AUG | SEP | OCT | NOV | DEC |

Green June Beetle Life Cycle in North Carolina (redrawn from USDA).

26

grubs deposit soil around the burrow entrance, making small mounds 2 to 3 inches in diameter which resemble ant mounds or earthworm castings. The grubs may also emerge and crawl about on their backs during warm, wet evenings. This is especially common after a warm afternoon thunderstorm.

As winter weather reduces soil temperature, the grubs dig deeper and remain inactive at the bottom of the burrow. These grubs may become active any time the soil temperature rises.

Larval digging may occur anytime during the late summer, through winter and into spring. By late May and early June the grubs mature and construct an earthen pupal cell glued together with a secretion. In about three weeks, adults break out of the pupal cell and dig to the surface.

After a fall rain, green June beetle larvae commonly push up mounds of soil.

Green June beetle grub in overwintering burrow (largest grub in middle), **Japanese beetle grubs** *(right and left).*

(left to right) **Green June beetle pupal cell, prepupa and pupa in opened cells.**

Black Turfgrass Ataenius
and **Aphodius**

The black turfgrass ataenius (BTA) has been collected from at least 40 of the 48 contiguous states and probably occurs in all 48. It is also known as the black fairway beetle in Canada. The first record of damage was reported from Minnesota in 1932. The next report of damage was in 1969 on two golf courses in New York, and in 1973 in Ohio. Since then, damage has occurred in most of the states where cool-season grasses are grown.

The adult BTA is a shiny black beetle, 3/16-inch (3.6 to 5.5 mm) long and 1/8-inch (1.7 to 2.4 mm) wide. The thorax has small pits scattered over the top surface and the wing covers have distinct longitudinal grooves. Newly emerged adults are reddish to dark chestnut brown but these become black in a few days.

Another insect, similar in general appearance and biology to BTA, but in the genus **Aphodius**, is also known to have damaged golf course turf in Colorado, Iowa, Michigan and perhaps other Mid-Western and North Central States.

BTA and *Aphodius* larvae look like small typical white grubs. BTA has one or two generations per year depending on location, and *Aphodius* apparently has one. BTA and Aphodius adults and larvae can be distinguished with a 10X hand lens.

Diagnosis. BTA and *Aphodius* larvae are known to feed in the root systems of annual bluegrass, bentgrass and Kentucky bluegrass. While most common on golf courses, BTA adults and larvae can be found in home lawns but damage is uncommon.

First symptoms of injury may appear from mid-June to mid-July when **turf shows evidence of wilting, similar to that caused by localized dry spots**. Wilted areas are especially visible when viewed toward the sun. Under continued stress from summer heat and larvae consuming roots, the turfgrass dies in irregular patches. Damage may reoccur in late August and early

(left to right) **Grass blade, BTA thrid instar larva, pupa, newly emerged "tineral" adult, older adult.**

September in areas where two generations occur. Populations of 250 to 300 larvae/ft² are frequently associated with heavy damage. Turfgrass areas under stress from heat, moisture shortage, or heavy traffic show the most severe damage.

The C-shaped white grubs are very small but third instars can be separated from other grubs using a 10X hand lens. The tip of the BTA abdomen has two distinct white anal pads and the few raster bristles are scattered at random. Smaller grubs have these characters but a microscope may be needed to see them. Full grown larvae are approximately 1/4-inch long.

Third instar *Aphodius* grubs can be distinguished from BTA by a **short V-shaped raster pattern** and a cleft anal pad at the tip of the abdomen. The V is visible with a 10X hand lens, but a dissecting microscope is required to see the cleft pad.

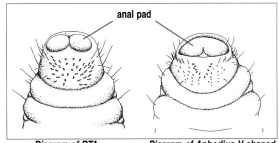

anal pad

| Diagram of BTA raster & anal pads. | Diagram of Aphodius V-shaped raster & cleft anal pad. |

27

First symptoms of damaging populations of BTA or Aphodius larvae on fairways can be irregular patches of wilt (upper left) that persist even with irrigation, and/or yellowing on greens and collars (lower left). Turf in these areas soon dies (upper and lower right).

Life Cycle and Habits. There are many species of *Ataenius* in the United States and most are dung feeders. *A. spretulus* and *A. strigatus* larvae feed on decaying humus (thatch) as well as living plant roots. Optimum habitat for BTA seems to be short cut turf with a moist compacted thatch layer.

BTA adults overwinter 1 to 2 inches below the soil surface, under leaves and other debris along the edges of fairways and in adjacent wooded areas. Adults begin to emerge and return to golf courses or other nearby expanses of turfgrass in late March when crocus and red bud are in bloom. Migration from overwintering sites continues through April and early May when adults often are seen on greens and swarming in flight over fairways on warm afternoons. Adults alight on the grass and immediately burrow into the turf.

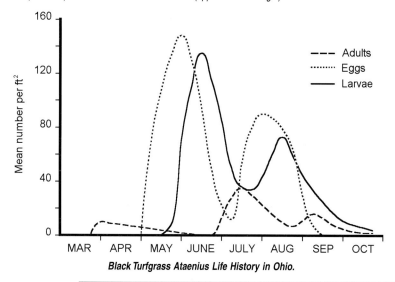

Black Turfgrass Ataenius Life History in Ohio.

In Ohio, **egg laying begins in early May when Vanhoutte spirea comes into full bloom** and continues into mid-June. Eggs are deposited in clusters of 11 to 12 in the soil or thatch just above the soil. Eggs hatch and larvae are present in thatch and soil from late May to mid-July. While the larvae feed on roots of bentgrass and Kentucky bluegrass, damage is most common on irrigated fairways with a high percentage of annual bluegrass. Symptoms of injury at this time include wilting of turf in spite of regular irrigation.

In late June and July, mature larvae burrow 1 to 3 inches into the soil, pupate, and become adults that emerge in July and early August. The reddish and black adults are often numerous beneath turfgrass killed by larvae. Black adults are seen around lights at night.

First generation BTA egg laying begins when Vanhoutte spirea comes into full bloom (left) and second generation egg laying begins at full bloom of Rose-of-Sharon bloom (right).

Where two generations occur, first generation **adults begin egg laying when Rose of Sharon comes into bloom** in July and produce a second generation of larvae. Second generation larvae can cause the same symptoms and significant injury as the first generation. <u>In Minnesota and other more northerly states, the second generation of BTA apparently does not occur.</u> In Ohio, completion of larval development and pupation of the second generation occurs in late August and September. These adults emerge and leave the fairways for overwintering sites during September and October. <u>Larvae that do not complete development to adults do not survive the winter.</u>

Aphodius. *Aphodius* larvae have been found damaging turfgrass on golf courses, especially in northern (Michigan) and western states (Colorado, Iowa). The <u>small black beetles have the shiny bodies</u> and elongate shape characteristic of BTA, but have been identified to be in the genus *Aphodius*.

A. granarius is the species most commonly found. The easiest method of distinguishing BTA adults from *Aphodius* is examination of the hind leg. The **hind leg of Aphodius has two bumps or spurs on the tibia** (the long segment just before the small series of ending tarsal segments). **Ataenius adults have a smooth tibia**.

Aphodius appears to have only one generation per year and <u>larval damage often occurs 2 to 4 weeks before first generation *Ataenius* damage</u>. Third instar larvae can be <u>distinguished from BTA by a short V-shaped raster</u> and a cleft anal pad at the tip of the abdomen. The V is visible with a 10X hand lens but the cleft pad is not.

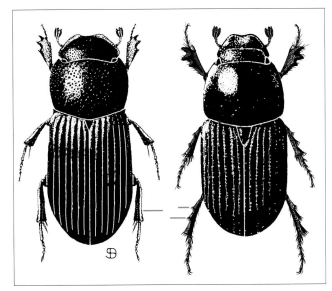

BTA adult (left) - note smooth hind tibia with no spurs. Aphodius adult (right) - note tibial spurs.

May or June Beetles, *Phyllophaga*

Over 150 species of May or June beetles (=Junebugs), *Phyllophaga* spp., are known in North America, but of these, only about 25 have been found attacking turfgrasses.

Phyllophaga are native to the United States and Canada. Some species, such as P. *hirticola*, P. *crenulata*, P. *tristis* and P. *ephilida*, are generally distributed from the Great Plains east. P. *crinita* is an important pest in Texas and Oklahoma and can also be found from Louisiana to Georgia. P. *latifrons* commonly attacks St. Augustinegrass in Florida. Only a few species like P. *anxia* and P. *fervida* are found all across North America.

<u>Adults of different species all have the same general shape but may differ considerably in size and coloration.</u> An expert is needed for species identification and usually only males can be used for accurate identification. Adults may be a light ochre brown as in P. *crinita* to almost black as in P. *anxia*. Adults are usually 1/2- to 1-inch (12 to 25 mm) long depending on species, and feed on the foliage of many shrubs and trees, such as oak, hickory, walnut, elm and poplar.

In general, May-June beetle larvae will feed on the roots of many types of plants including turfgrasses, pasture grasses, field crops, trees, shrubs and flowers.

Diagnosis. In recent years, this group has become less severe as damaging pests in the northeastern states, probably because of dominance by imported grub pests such as the Japanese beetle. However, in regions of Canada where other grub species are not common, *Phyllophaga* species are the primary grub pest. In southern and western states, *Phyllophaga* are often present in high populations. Damaged turf wilts as if under drought stress and eventually this turf dies in irregular patches. Some southern grasses

photo: J. Polivka

Typical May-June beetle adult.

Pin oak nearly defoliated by May-June beetle adult feeding.

may not show these symptoms as readily because of moist sandy soil and deep roots. However, digging by mammals in search of the grubs as food can cause serious damage.

Specific identification of May-June beetle grubs is very difficult and requires an expert. However, almost all larvae have a broadly **V-shaped anus and a raster with two parallel rows of bristles pointing towards each other at either end**. Full grown grubs are about 7/8 to 1-1/4 inch (20 to 30 mm) long.

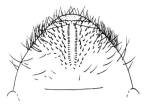
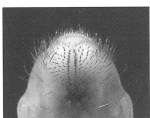

Diagram of May-June beetle raster.   *Photo of May-June beetle raster.*

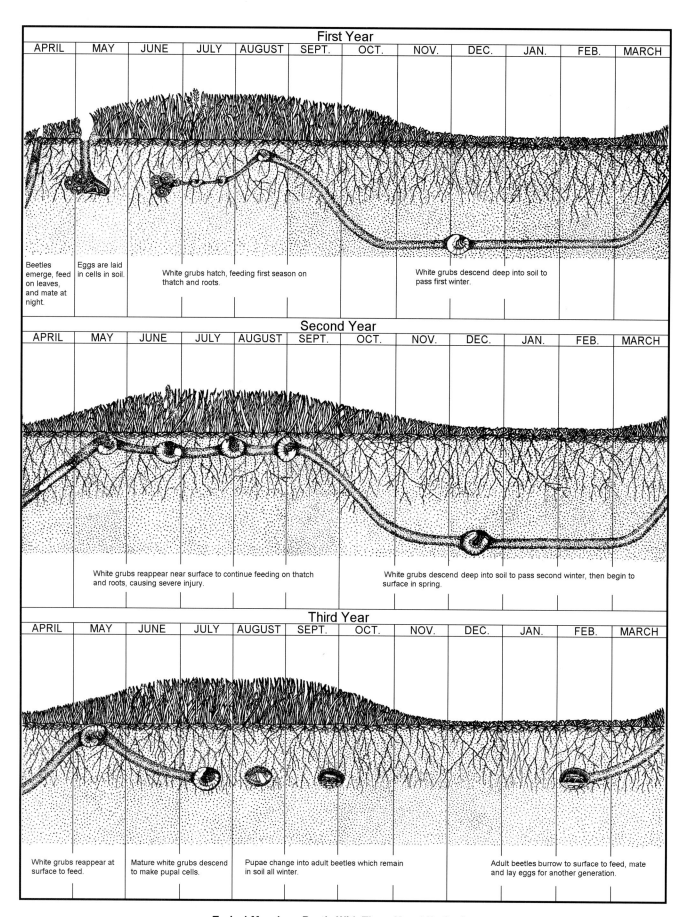

First Year

| APRIL | MAY | JUNE | JULY | AUGUST | SEPT. | OCT. | NOV. | DEC. | JAN. | FEB. | MARCH |

Beetles emerge, feed on leaves, and mate at night.

Eggs are laid in cells in soil.

White grubs hatch, feeding first season on thatch and roots.

White grubs descend deep into soil to pass first winter.

Second Year

| APRIL | MAY | JUNE | JULY | AUGUST | SEPT. | OCT. | NOV. | DEC. | JAN. | FEB. | MARCH |

White grubs reappear near surface to continue feeding on thatch and roots, causing severe injury.

White grubs descend deep into soil to pass second winter, then begin to surface in spring.

Third Year

| APRIL | MAY | JUNE | JULY | AUGUST | SEPT. | OCT. | NOV. | DEC. | JAN. | FEB. | MARCH |

White grubs reappear at surface to feed.

Mature white grubs descend to make pupal cells.

Pupae change into adult beetles which remain in soil all winter.

Adult beetles burrow to surface to feed, mate and lay eggs for another generation.

Typical May-June Beetle With Three-Year Life Cycle.

Life Cycle and Habits. The life stages of May-June beetles are typical of white grubs, however, depending on the species, 1 to 5 years are required to complete a generation. Species that attack cool season turf take 3 to 5 years to complete development, usually three years in New York across to Nebraska and south. Species in the middle states and south (transition and southern turf zones) may take 1 to 2 years to develop. *P. crinita,* a common Southwestern pest, takes only one year to develop, and *P. latifrons* in Florida takes only one year. In northern states and Canada, adults begin to emerge in May and June, thus their name "May beetles" and "Junebugs."

Adults emerge from the soil at dusk and fly to trees and shrubs and feed on foliage. If the evening temperatures are high enough, adults continue feeding until dawn when they burrow into the soil.

High populations can severely defoliate preferred plants such as oaks and hickories. The adults are also highly attracted to night lights and often alarm people with their buzzing flight.

After feeding several nights and mating, females burrow into moist soil to a depth of 2 to 6 inches where they lay 20 to 30 eggs individually packed into balls of soil. These eggs must absorb moisture from the surrounding soil to develop. After 20 to 40 days, the eggs hatch and the young larvae burrow upwards in search of organic matter, thatch, and plant roots. The three-year species remain as larvae for two years and do not pupate until the third summer. Pupae transform to the adults by early fall, but these do not emerge until the following May or June. In the far northern states, some species may require 3 to 4 summers to complete larval development before they pupate.

Mole Crickets in General

Seven species of mole crickets may be found in North American turfgrass, but only four of these are considered important pests. The tawny mole cricket (previously missidentified as the Changa or Puerto Rican mole cricket) and the southern mole cricket are the most damaging species. The short-winged mole cricket occurs occasionally at pest levels. These three species were introduced from South America. The native mole cricket is considered a nuisance pest in the northern part of its range.

Mole crickets that damage turf are found south of a line running from mid-North Carolina through mid-Louisiana and into southeastern Texas. Damage consistently occurs south of a line running from southeastern Texas along the Gulf States into southeastern North Carolina where southern grasses, such as centipedegrass, bahiagrass, bermudagrass and St. Augustinegrass, are affected. The native mole cricket occurs throughout the eastern half of the United States. While the tawny and southern mole crickets are the most common and damaging pests, shortwinged and native species have also been known to cause damage in certain areas.

Diagnosis. Mole crickets tunnel through soil like their mammal counterpart. This tunneling breaks up the soil around turf roots and the turfgrass often dies due to desiccation. The tunneling trails themselves are considered unsightly and interfere with ball roll on golf course greens and fairways. During mating and overwintering, adults often push up mounds of soil around their permanent burrows. At this time, the adults are not feeding extensively enough to kill large patches of turf. Severe damage occurs in summer months when nymphs are actively feeding on turfgrass roots. Heavy

Mole cricket damage to bermudagrass. Note streaked damage, soil mounds and emergence holes that are characteristic of mole cricket activity.

infestations during this period may result in large dead patches and exposed soil. St. Augustinegrass does not show severe symptoms of damage from mole crickets, possibly because of its dense growth habit and fibrous root system.

Sampling. A soap solution of liquid detergent (two tablespoons of liquid Joy® dish washing detergent in two gallons of water) may be used for ***mapping areas of mole cricket infestations*** and determining stages present. The solution is poured over areas of activity (tunneling or mounding). Mole crickets come to the surface within a few minutes after application. This system works best when applied after rain or irrigation which brings the crickets closer to the surface (their burrows may be 1 to 2 feet in the soil). In our experience, this solution has not damaged turf, but it is always wise to test the solution in an inconspicuous area to determine if damage may occur.

Sampling to determine if control will be needed is best done as soon as spring (March to May) movement is noticed (major flights at night or suddenly increased tunneling and soil mounding).

pronotal patterns can be used to identify species

wings

head

forelegs can be used to identify species.

Male tawny mole cricket with major body parts used to identify species.

General Life Cycle and Habits.

Mole crickets undergo incomplete (gradual) metamorphosis with egg, nymph and adult stages. <u>Nymphs resemble adults but are smaller</u> and do not have developed wings. Most species have a single generation per year with adults maturing by fall, but do not lay eggs until the following spring.

Tawny mole cricket life stages (left to right): **nymphal instars 1 through 8, adult female & adult male.**

Mole Cricket Identification

Four common species of mole crickets can be identified by examining the claws (dactyls) on the front leg.

The northern mole cricket has four major claws while the imported species have two.

The space between the tibial dactyls, either V-shaped or U-shaped separates the tawny mole cricket from the southern and shortwinged.

Color patterns on the pronotum are also useful in separating species.

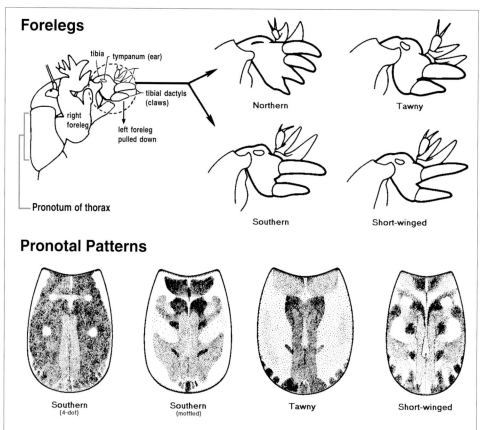

Forelegs

tibia
tympanum (ear)
tibial dactyls (claws)
right foreleg
left foreleg pulled down
Pronotum of thorax

Northern
Tawny
Southern
Short-winged

Pronotal Patterns

Southern (4-dot)
Southern (mottled)
Tawny
Short-winged

(illustrations redrawn from Walker et al. 1984, Glorida Agr. Exp. Sta.).

Tawny Mole Cricket

The tawny mole cricket was introduced into Georgia around 1899; now found throughout Florida and south of a line running from the southern tip of North Carolina, across Georgia and to the southern tip of Alabama. This species is a native of coastal South America from northern Argentina through Brazil and in Columbia, Panama and Costa Rica. It prefers to live in bahiagrass but damages close cut bermudagrass and occasionally St. Augustinegrass, centipedegrass and zoysiagrass.

Fully grown <u>adults</u> are about 1-1/4- inch (30 to 34 mm) long and 3/8-inch (8 to 10 mm) wide. They are a light <u>tawny brown</u> and have the obviously modified front legs and enlarged thorax typical of mole crickets. The forewings are shorter than the abdomen and males have a darker rasp (the sound producing organ) at the forewing base. <u>The V-shaped space between the tibial claws (dactyls) is a species identification characteristic</u>.

***Tawny mole cricket adults, male** (left - note dark mark on upper wing base)* **and** *female (right).*

32

Diagnosis. Tawny mole crickets produce typical mole cricket damage to turf. **This species prefers to feed on roots** but also damages turf by tunneling and pushing up mounds of soil around permanent burrows. Loosened and uprooted turf usually withers and dies, leaving trails of dead turf. Heavy populations can completely kill turf, leaving bare patches of soil.

To determine which species are present, capture nymphs or adults by using a soap drench and inspect the front leg dactyls with a 10X hand lens for positive identification.

Life Cycle and Habits. Male mole crickets locate preferred habitats in the spring and call with their *toad-like trill* from the entrance of their burrows. This occurs for about an hour shortly after sunset. Mole crickets can fly more than six miles in a night and may fly more than once. Females are attracted to the calls of males in the spring when temperature and moisture allow easy movement and flight. Egg laying may begin in March, but 75% of the eggs are laid between May 1 and June 15. During the mating and oviposition periods, the adults are extremely active, emerging from the soil at dusk to crawl about, tunnel and fly. Flight in mass commonly occurs after rainfall and adults are attracted to lights.

Mated females burrow a few inches into suitable soil and construct an egg chamber in which they lay an average of 35 eggs. Females may construct three to five chambers and lay 100 to 150 eggs total. The eggs hatch in about 20 days and the young nymphs

Diagram of tawny mole cricket front leg. Note narrow, V-shape between the two large claws (dactyls).

Photo of tawny mole cricket front leg.

burrow upward to seek food. Nymphs prefer to feed on plant roots but will eat small insects and can be cannibalistic. Larger nymphs will often attack smaller nymphs and eggs. Nymphs continue to feed, molt and grow through the summer months. By the end of October, 85% of the nymphs have become adults. The remaining nymphs continue development very slowly and most mature by the following spring.

Adult and nymphal behavior are highly regulated by temperature and soil moisture. Most feeding occurs at night, especially after rain showers or irrigation during warm weather. These insects may tunnel 20 feet per night in moist soil. During the day, individuals tend to return to a permanent burrow which is deeper in the ground. The crickets may remain in these permanent burrows for considerable periods during cool winter temperatures or dry spells.

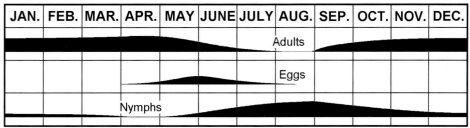

JAN.	FEB.	MAR.	APR.	MAY	JUNE	JULY	AUG.	SEP.	OCT.	NOV.	DEC.

Adults

Eggs

Nymphs

Tawny Mole Cricket Life Cycle in Northern Florida.

Southern Mole Cricket

Southern mole crickets were introduced into Georgia at the turn of the century and are now found south of a line running from mid-North Carolina through mid-Louisiana and into eastern Texas. It is probably a native of South America where similar mole crickets have been found. This species also occurs in northern Argentina, Uruguay, Paraguay, Bolivia and Brazil.

Apparently southern mole crickets prefer to **prey on other insects and each other**. However, considerable feeding on plants occurs; 41% were found to have plant material in the gut. It is often found in bahiagrass, bermudagrass and occasionally St. Augustinegrass. Centipedegrass and zoysiagrass are rarely damaged.

Adults are about 1-1/4-inch (32mm) long and 3/8-inch (9 mm) wide, are reddish to dark brown and have either distinct pale spots on the pronotum or a general mottled color. The forewings are shorter than the abdomen, and the hind wings extend just beyond the tip of the abdomen. A U-shaped area between the tibial dactyls separates this species from the tawny mole cricket but not the shortwinged. Males also have a darker rasp and file (sound producing organ) at the base of the forewings.

Southern mole cricket female (left) and male (right). Note four pale spots on pronotum, a field character used for identification.

Diagram of southern mole cricket front leg showing U-shaped space between claws (dactyls)

Photo of southern mole cricket front leg.

Diagnosis. The southern mole cricket produces typical mole cricket damage (turf death in strips, soil mounds), though most damage is due to tunneling. Nymphs and adults tunnel and push up mounds of soil. Uprooted turf soon wilts and dies from desiccation.

Capture nymphs or adults by using a soap drench and inspect the front leg dactyls with a 10X hand lens for positive identification.

Life Cycle and Habits. This species has a cycle similar to the tawny mole cricket. Egg laying occurs from mid-March into June, but some adults continue laying eggs into September. The nymphs mature rapidly through the summer, but only about 25% reach adulthood by winter. Some males call during the fall and mating may occur at this time. Nymphs that do not complete development by fall develop slowly during the winter (December to March) with peak

adult populations appearing in May. Nymphs and adults of this species have been found during all months of the year.

Most activity is noticed after rain or irrigation. A spring flight period occurs, and calling, mating and egg laying takes place. Males produce a **bell-like trill** for an hour or longer just after sunset. In south Florida, males may call anytime the temperature reaches 60°F.

The southern mole cricket is more of an omnivore (feeds on plant and animal matter) than the tawny mole cricket. Only 5% have been found with plant material alone in the gut while 59% had only animal material in the gut and the rest had a combination. Thus, it is suspected that turfgrass damage caused by the southern mole cricket is due to its tunneling and soil mounding rather than actual root feeding.

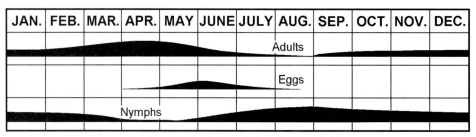

JAN.	FEB.	MAR.	APR.	MAY	JUNE	JULY	AUG.	SEP.	OCT.	NOV.	DEC.
							Adults				
							Eggs				
		Nymphs									

Southern Mole Cricket Life Cycle in Northern Florida.

Shortwinged Mole Cricket

The shortwinged mole cricket occurs in localized areas of Florida and Georgia in the United States and is native to Argentina, Paraguay and Brazil. It has also been introduced into Puerto Rico, Virgin Islands, Cuba, Nassau and Haiti. This species prefers plant material as food.

Adults are similar to other mole crickets except that the hind wings extend no more than one-third the length of the abdomen. The hind wings are very small and the ocelli (the pair of small single eyes between the larger compound eyes) are smaller than those of the southern mole cricket. This pest **can not fly** and is most likely moved about in soil that is hauled from infested to uninfested areas.

Diagnosis. The shortwinged mole cricket produces typical mole cricket damage but the damage is often attributed to the nymphs of one of the other species. This pest does not fly and thus tunnels to disperse. The nymphs and adults are **difficult to distinguish from southern mole cricket nymphs** which also have a U-shaped space between the tibial dactyls. However, the ocelli (simple eyes between the compound eyes) are smaller in the shortwinged mole cricket.

Life Cycle and Habits. This species is probably the least understood of the turf infesting mole crickets. Apparently it was more of a pest before the tawny mole cricket was introduced. In recent years, however, more shortwinged mole crickets have been found and its distribution seems to be expanding, probably because it is often transported with soil. This species cannot fly and must naturally spread by crawling or tunneling. The males do not make a mating call but produce low chirping sounds when a prospective mate is encountered. The life cycle is similar to the southern mole cricket, in that most nymphs become adults by fall.

Shortwinged mole cricket adult.

Diagram of shortwing mole cricket front leg showing U-shaped space between claws (dactyls).

Photo of shortwing mole cricket front leg.

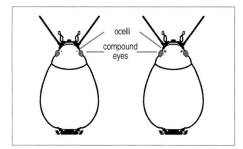

Pronotum and head of shortwing (left) and southern (right) mole crickets showing compound eyes (large red spots) and ocelli (small red spots). Shortwing mole crickets have smaller ocelli.

Native Mole Cricket

The native mole cricket (=northern mole cricket) occurs in North America, east of the Rocky Mountains. It is most common near wetland habitats (river flood plains, marshes and ponds). It has also been found in Michigan, Ohio and southern Ontario, Canada.

This species has habits similar to the southern mole cricket in that it seems to **prey on other insects**. The adults are about 1-1/4-inch (28 to 32 mm) long and 3/8-inch (8 to 10 mm) wide, darker brown than the imported species and the pronotum has no pattern. Adults have short front wings but the hind wings may extend just past the tip of the abdomen. Many adults have both sets of wings short and can not fly. Adults with normal hind wings can fly though they rarely do so except during the spring mating period.

Diagnosis. Native mole crickets rarely damage turf except when they tunnel under well maintained turf near their semiaquatic habitats. Golf course greens constructed near **river banks**, **wetlands** or **ponds** have been damaged. Damage usually appears as tunnels or pushed up mounds of soil. Loosened and uprooted turf withers and dies or is scalped when mowed.

Capture nymphs or adults by using a soap drench and inspect the front leg dactyls with a 10X hand lens for positive identification. Native mole crickets are the only species with four claws (dactyls) instead of the two claws found on tawny, southern and shortwing mole crickets.

Life Cycle and Habits. In early spring, overwintered males construct a calling chamber near the soil surface when temperatures begin to reach 60°F. From these burrows, a series of low pitched trills are made to attract females. After a warm spring rain, females and males may take flight. Both sexes can be attracted to lights and areas where males have established calling chambers.

Native mole cricket adult. Note that the forewings are short and may resemble shortwing mole crickets. Native mole crickets have a nearly uniform brown pronotum, a four-clawed (dactyls) front leg and no bands on the hind legs.

Diagram of native mole cricket front leg showing four claws (dactyls). *Photo of native mole cricket front leg.*

Mated females seek out habitats with a high water table for establishing permanent burrows and to begin egg laying. Favored sites are river flood plains, stream, pond and lake banks as well as marsh lands. Nymphs disperse by tunneling through the soil in search of insects and other invertebrates as food. This species is cannibalistic and will feed on other species of mole crickets.

The nymphs mature by early fall, but in northern states are not adults until the following spring. During periods of drought, this species digs deeper in the soil to reach the water table. During the winter months, it digs below the freeze line to survive.

Ground Pearls

Ground pearls are actually the immatures of a soil-inhabiting scale insect that completes the development inside a protective pearl colored shell (cyst) which it secretes. Warm-season grasses such as bermudagrass, centipedegrass, zoysiagrass and St. Augustinegrass from southern California to North Carolina are damaged by this pest. Damage is most common in centipedegrass and bermudagrass.

Diagnosis. Irregular patches of turf appear unthrifty, and over a year or two, thin out or die. Removal of plant fluids and probable injection of salivary substances causes the turf to turn yellow then brown in irregular patches. Damage is most common during dry spells.

Life Cycle and Habits. Because these insects are essentially subterranean for most of their lives, little is known about their exact behavior. Generally it is thought that there is only one generation per year.

The adult females are pinkish sack-like forms about 1/8-inch (2 to 3 mm) long, and have well developed front legs and shorter

photo: P. Cobb

Centipede grass lawn damaged from ground pearls. With increased fertility and regular irrigation, this lawn dramagically improved within a year.

second and third legs. Mature females occur in mid- to late May, during which time they emerge from their hard waxy cysts, burrow to the soil surface and mate with the rarely seen, gnat-like winged males. Once mated, the females dig 2 to 3 inches into the soil and form a new waxy coat in which they lay about 100 eggs.

The eggs hatch into crawlers (1/100-inch long) which disperse in the soil in search of grass roots. When suitable grass roots are found, the crawlers insert their piercing mouthparts, attach themselves and secrete a hard, yellow to light purple, waxy coating (=cyst) which is the ground pearl stage. The nymphs continue to develop inside this cyst and overwinter attached to the root.

There is some evidence that a small portion of the cysts can remain dormant in the soil for several years.

Ground pearl cysts in soil and around roots of centipedegrass.

Ground pearl cysts (left), empty cyst shells (right), and adult female (bottom).

The Ants

More than 600 species of ants occur in North America, some of which inhabit home lawns, golf courses and other turf areas. **Generally, ants are beneficial insects, feeding on pest insect eggs, other insects and whatever they can find**.

Though various ant species occur in turf from time to time, two have caused most of the problems - fire ants and the "turfgrass ant." The red imported fire ant was introduced from South America in 1930 and now infests most of the southern states from Oklahoma to North Carolina. In addition to its mound building habits, this ant's stinging and biting behavior against anything that disturbs their mounds is a serious problem. Venom injected from stings causes painful lesions that are slow to heal and may cause allergic reactions.

The **turfgrass ant**, *Lasius neoniger*, probably occurs over most of the United States. Though beneficial because it feeds on insect eggs and small larvae, this ant is also known for building mounds on golf course tees, greens and fairways which damage turf and mowing equipment.

Other nuisance ants include the leafcutter, harvester and Allegheny mound ants. The leafcutter ant is most common in Texas and Louisiana where it can make large mounds near trees. It often clears trails through turf in order to carry back pieces of tree leaves. Harvester ants are also associated with the arid southwest and the

Leafcutter ant worker.

Harvester ants feeding on crushed snail.

foraging ants also have the habit of clearing trails through turf. The Allegheny mound ants are most common in the northeastern states and they can build large mounds in turf, usually near wood lots. These mounds serve as solar collectors to heat ant larvae and pupae, thereby increasing their developmental rate.

Harvester ants collect seeds of grasses and other plants to store as food in their nests. They have the habit of clearing away all vegetation around their nest opening and create cleared trails in turf.

Leafcutter ant trail extending into a bermudagrass fairway from nest opening at base of tree.

Velvet ants are not really ants, but a type of wasp whose females usually have no wings. They commonly parasitize grasshopper egg clusters and are frequently found in southern states. The females can inflict a painful sting if provoked.

"Turfgrass Ant"

This species is known as a pest on golf courses from Quebec to Florida, Idaho to Colorado and probably most of the United States. While generally considered a nuisance scavenger, this small brown ant builds 3- to 5-inch diameter mounds of fine sand and soil on tees, greens, fairways and other aesthetically sensitive areas. Frequent mowing flattens the mounds causing the turf to die in small circles. Mounds also dull the blades of mowing equipment. The ant itself does not damage turf and it is not known as a pest of home lawns.

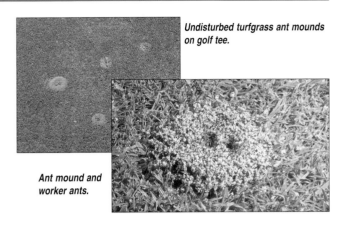

Undisturbed turfgrass ant mounds on golf tee.

Ant mound and worker ants.

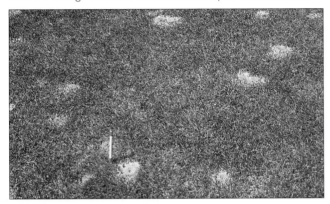

The soil brought up by the turfgrass ant commonly covers and kills the short cut grass of golf greens, tees and fairways.

Life Cycle and Habits. Each colony has multiple subdivisions and many entrances but only one queen which lays eggs that for most of the year produce infertile female workers. In the summer, the queen lays eggs that produce winged males, and others that develop into winged, reproductive females or queens.

In late August, large numbers of the winged forms emerge (swarm) from the colony in a mating flight. After mating, the males die but the females chew off their wings, dig into the soil and develop new nests in golf course fairways or other open grass areas. The first eggs laid by the new queen develop into small workers that collect food for the queen who lays more eggs that develop into normal workers.

The ants feed on dead insects, insect eggs, earthworms and any other acceptable food. The food is regurgitated and fed to the queen, young workers and larvae. These ants also collect certain species of aphids, carry them underground and place them on turf roots. The aphids are carefully tended and produce a sweet fluid called honeydew that the ants use as food.

The ants move deeper into the soil in late fall and resume surface activity in early spring (late April). Burrows may extend three feet or more into the soil.

Fire Ants

Four economically important species of fire ants are found in the United States. They are: native fire ant, red imported fire ant, black imported fire ant, and southern fire ant.

The red imported fire ant, the most important species, is a native of Brazil but was introduced into Alabama between 1933 and 1945. This pest now inhabits the area south of a line running from mid-North Carolina across to the Dallas-Ft. Worth, Texas area and down to Corpus Christi. The black imported fire ant is a native of Argentina and Uruguay which was imported into the Mobile, Alabama area as early as 1918. This ant is still localized in the area were Mississippi and Alabama join. The native fire ant was originally located across the southern states but has generally been displaced by

the red imported fire ant. The southern fire ant is a native to Arkansas, Oklahoma and Texas.

Diagnosis. Fire ants do not attack turf but cause problems when they build earthen mounds to warm eggs, larvae and pupae. They have a notorious sting which may cause a burning and itching sensation at minimum and serious welts or allergic shock at its maximum. Unsuspecting people, pets, and cattle may be severely injured when the mounds are accidentally disturbed.

The rounded cone shaped mounds of the red imported fire ant may be several inches to two foot in diameter and an inch to eight inches tall.

Fire ants usually grip the skin with their mandibles (mouthparts) and inject a toxin with their abdominal stinger.

When a fire ant mound is disturbed, hundreds of ants emerge from the broken surface.

Red imported fire ant workers.

establish a new colony by digging a small hole in the soil and closing up the entrance. Inside this chamber, the queen lays 15 to 20 eggs in 2 to 3 days. More eggs are added over the next week by which time the first eggs hatch. The queen picks up the young larvae and sorts them into groups. The larvae are fed a liquid regurgitated by the queen.

After 20 to 25 days, the larvae pupate and tiny workers emerge 4 to 7 days later. These first workers are about 1/5 the size of the smallest workers found in an older colony. These workers break open the nesting chamber and begin foraging for insect food and start to enlarge the nest. The queen, now fed by workers, begins to lay more eggs, which are cared for by the workers. If food and water are adequate, the colony steadily grows over the next few months. If a colony is established in June, it may contain 6,000 to 7,000 individuals by the following December.

As the soil temperature drops, the colony growth slows. By the following June, a one-year-old colony may have 10,000 to 15,000 workers and can be producing new winged forms. Colonies 2- to 3-years-old may have 20,000 to 200,000 workers. Established mounds will have a central pile of granular soil with openings and often smaller mounds around the perimeter.

Depending on the species and locality, fire ants may be yellow to reddish brown to black in color. An expert should be consulted if species determination is necessary.

Life Cycle and Habits. Established colonies produce new queens and winged males during warm spring and summer months. These winged reproductives swarm periodically, usually 5 to 9 times a year, often after a rain. Mated queens attempt to

Colonies may move the mounds in search of food, when regularly disturbed by mowing or when pesticides are applied.

Cicada Killer

This large wasp is found in all the states east of the Rocky Mountains but is most common in states where annual cicadas are prevalent. Neither the adult nor other life stages alone damage turf. However, because of the adult habit of creating mounds of soil, it is sometimes considered a pest of golf courses and other areas. Though not as aggressive as other wasps in defending their nesting sites, their size and the buzzing sound they make frightens people, especially golfers who find themselves in the area of their burrows. **Females can inflict a painful sting, but do so only when handled or severely provoked**.

The large adult wasps have rusty red heads and thoracic areas. The wings are tinted with orange and the abdomen is banded with black and yellow. The adults may be 1-3/16 to 1-3/4 inch (30 to 45 mm) long and have a wingspan of 2-3/8 to 4-inches (60 to 100 mm).

Cicada killer females hunt cicadas, inflict a paralyzing sting, grasp their prey with their legs and return to the burrow.

Diagnosis. Wasps dig burrows and create mounds of soil around the entrance which is approximately 1/2 inch (12 mm) in diameter. Males are very protective of territory and often buzz or dive on people who enter the burrow territory being protected.

Life Cycle and Habits. This insect overwinters as a prepupa in a cocoon located 7 to 20 or more inches under the surface. When spring temperatures warm the soil, the pupal stage is formed and the adults burrow to the surface in June and July. Males tend to emerge first and establish flight territories, usually where females are to emerge and dig burrows. Males fight off other intruding males and buzz at any moving object, including people, in the area. Fortunately, **males have no sting**, but unsuspecting people are often shocked by the loud buzzing and their attack flight activities. Males may actually strike peoples' heads and backs.

Cicada killer male surveying territory.

The females are quite docile and after mating are occupied with constructing a burrow. Burrows are dug straight into the soil, are angled slightly and may extend three or more feet into the soil. At the end of each tunnel, secondary tunnels are dug that end in a chamber. The females search tree trunks and limbs for annual cicadas only. When located, the female inflicts a paralyzing sting, grasps the prey with her legs and flies back to the burrow (considering the weight of the cicada, this is an amazing feat!). Each chamber is provided with one to three cicadas. An egg is laid on the cicada, it is placed in the chamber and the chamber is sealed. Additional chambers are constructed until cicadas are no longer available.

The wasp egg hatches in 2 to 3 days and the voracious larva quickly devours the inside of the paralyzed cicada. Only the cicada exoskeleton remains. The larvae mature by fall, spin a cocoon, shrink and prepare to overwinter. A single generation occurs each year.

Cicada killers commonly nest in groups in or around sand traps.

Crane Flies or Leather Jackets

Crane flies or leather jackets are common names for large mosquito-like flies and their tough skinned larvae. Crane fly larvae live in aquatic, semiaquatic and terrestrial habitats. The **European crane fly** is considered a turfgrass pest in British Columbia, Nova Scotia, Washington and Oregon. Other species occur in the cool-season turf areas of North America and are generally considered nuisances more than pests.

Adult crane flies are slender bodied, brownish-tan, long-legged, mosquito-like flies with two smoky-brown wings. They are commonly seen flying singly or in swarms around ponds, streams, meadows and golf courses at twilight. European crane fly adults emerge after sunset and in southwestern British Columbia and western Washington, can become so numerous as to cover the sides of houses overnight. On golf courses, the adults of other species become nuisances as they fly in the faces of golfers passing through swarms.

Diagnosis. Larvae feed on roots and crowns below the surface and on grass blades and stems on the surface. The **European crane fly causes bare and/or sparse turf areas** which become evident from March to April.

Unless extremely numerous, other species apparently cause little damage. Native species also feed on decaying thatch or may be associated with turf damaged during winter months by diseases.

Life Cycle and Habits. The European crane fly seems to require mild winter temperatures, cool summers and average annual rainfall of at least 24 inches. Adult flies emerge from lawns, golf courses, pastures and roadsides from late August to mid-September. Mated females begin to lay black, oval eggs within 24 hours after emerging. In about two weeks the eggs hatch into small, brownish maggots which begin feeding by using their rasping mouthparts on plant roots, rhizomes and foliage. By winter the larva has molted twice and reached the third instar. These larvae feed slowly during winter temperatures and reach the fourth instar in April and May.

The leather jackets stay underground during the day but come to the surface to feed on damp, warm nights. Damage resembling that caused by black cutworms can occur on golf course greens.

Crane fly larva. Note black head and characteristic fingerlike projections around tip of abdomen.

photo: J. Law

European crane fly larvae on surface after an insecticide application, and characteristic sparse turf caused by larval feeding.

Crane fly larvae (=leather-jackets).

Cranefly adult on wall.

Crane fly adults often congregate on sides of buildings. Though they look like giant mosquitoes, they are harmless

Notes

Chapter 4

Principles of Controlling Soil-Inhabiting Pests

Japanese beetle grubs at the soil-thatch interface, a difficult TARGET to reach with control agents - chemical or biological.

The Target Zone

The **Target Zone** for control of most soil-inhabiting insects is the first one to two inches of soil. However, in both cool- and warm-season turfs, the soil may be covered by a layer of thatch through which the control materials must pass before reaching the target zone. Thatch serves as a binding site for insecticides and a **difficult barrier** for some biological agents to pass through. The addition of soaps or other wetting agents appears to have little affect on reducing the binding potential of soil or organic matter. The control material **must reach the target zone in the proper concentration to have the desired affect**. The degree to which this is achieved is directly related to the degree of control achievable.

Mobility is more readily achieved when thatch is thin, loose or not present. However, in long established turf sites, the constant dying of plant roots, stolons, rhizomes and crowns results in an accumulation of organic matter in the first two inches of soil (commonly 10% or more) which also serves as a binding site. While the organic matter content of thatch (often greater than 30%) and soil impede mobility, it also provides a protective filter too slow to stop more extensive mobility past the target zone.

Even when there is no thatch, through management or natural decomposition, the first two inches of soil in established turf commonly contains at least 10% organic matter due to the constant dying of plant parts. The binding of insecticides to this zone or thatch is a mixed blessing. While having a major influence on reducing the potential for ground water contamination, binding in this zone provides a reservoir of insecticide residues for pests such as grubs or mole crickets to contact and consume as they feed.

Feeding Habits

Knowing the feeding habits of soil-inhabiting insects is essential to understanding how and why control is or is not achieved. The primary means by which control agents enter the body of the target insect is through the natural openings (mouth, anus, spiracles) or

The Target Principle

Materials directed at controlling damage from soil-inhabiting insects must reach the primary feeding and/or activity zone (target zone) of the target insect(s) to be effective. Focusing on this objective when applying materials or employing other control strategies is to apply the **Target Principle**.

ingestion. Contact with a treated surface also occurs, however, with some exceptions, is generally secondary to the **impact of ingestion**.

What do grubs eat? The standard answer often is, "turf roots." This is an incomplete and actually incorrect answer. A more accurate answer would be, "**whatever is in front of them**." Grubs are incapable of feeding only on roots. Instead, they ingest the entire medium - roots, soil, soil organic matter - that occurs in their zone of habitation. Generally, this zone is the upper two inches of soil when no thatch is present and the upper one inch of soil when thatch is present. Soil inhabiting insects, such as mole crickets, consume plant and animal materials as well as soil particles. Turf inhabiting ants are general scavengers and predators, but they do not consume plant parts.

Masked chafer grubs exposed by pulling back the damaged turf. Notice the underside of the thatch (left) showing grub feeding burrows and tunnels.

Applied control agents must reach the feeding-activity zone (the Target Zone) of a soil-inhabiting pest to achieve control. The agent is adsorbed to varying degrees and distributed at and through the soil and thatch above the target zone. **The target insect ingests thatch and/or soil organic matter containing the agent, which is then absorbed by the insect as the food passes through the digestive system.** Living biological agents, such as insect parasitic nematodes, must wriggle through the thatch and/or soil and reach the target zone in sufficient numbers to find and infect the target pest.

Irrigation - Rain

Control materials vary widely in water solubility and capacity for adsorption to organic matter. Water, as rain or irrigation, does not completely circumvent adsorption, but it does accomplish as much movement in the target zone as possible. Generally, control materials for soil inhabiting insects should not be applied to very dry thatch or soil. Greater mobility is achieved when both thatch and soil are first moistened. In order to minimize ultraviolet (UV) degradation, hasten mobility and obtain a maximum effect, liquid materials should be irrigated in **immediately** after application.

Generally, granular materials should be applied when grass blades are dry so the particles bounce off the grass blades and sift deeply into the turf. While the urgency to irrigate is not as immediate as for liquid materials, it should be done as soon as possible.

Permanent, in-ground irrigation systems are excellent for helping move controls into the target zone, but portable lawn sprinklers can accomplish the same task.

With regular irrigation or rain, soil-inhabiting insects such as grubs usually remain in the target zone. However, if the surface soil dries, these insects may move deeper into the soil profile. Timely pre- and posttreatment irrigation often stimulates these insects to remain in the Target Zone.

When irrigation is not possible, applying control materials just before an impending rain can help move the materials to the target zone.

Subsurface Placement

Development of subsurface placement equipment that bypasses thatch and places the control agent directly into the Target Zone has been successful for control of mole crickets. This system has had low success and acceptance for control of soil-inhabiting pests in the cool season areas.

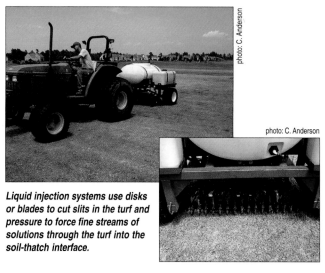

photo: C. Anderson

photo: C. Anderson

Liquid injection systems use disks or blades to cut slits in the turf and pressure to force fine streams of solutions through the turf into the soil-thatch interface.

42

Subsurface applicators that insert granules use disks or blades to slice grooves in the turf. Granules must be placed in the grooves <u>at the target zone</u> in order to achieve maximum control.

The mobility and feeding habits of mole crickets, low density of bermudagrass, and its ability to recover quickly after the slicing effect of application equipment are among the reasons for the <u>success and acceptability of this approach among southern golf course superintendents</u>. Since mole crickets move considerably, they often cross the areas in the turfgrass profile where the pesticide has been placed and are controlled by contacting and ingesting the toxicant.

Control of pests such as grubs in cool-season turfgrass requires that placement not go beyond the target zone. Equipment to consistently accomplish this has not yet been developed. Further, cool-season turfs are very dense and more difficult to slice into and <u>the visible effects of slicing may not be acceptable to golf courses</u>. High pressure injection has also had limited success.

In the authors' experience, despite precise placement into the Target Zone in strips two inches apart, <u>control of grubs has been inconsistent</u> and unacceptably low. The reason for this may be that grubs simply will not feed through (or avoid) the relatively high concentration of material placed into their activity zone and survive by remaining between the strips. Apparently, this does not occur when mole crickets are the target.

The recent labeling, effectiveness, low toxicity and long residual activity of new classes of insecticides have reduced interest in this approach to control of cool-season soil-inhabiting pests. Subsurface placement may have more application to installation of biological and other grub control agents in home lawns.

Chemical Control - Insecticides

Most of the insecticides currently labeled for control of soil-inhabiting insects have the ability to reach their target if used correctly (<u>according to their label instructions</u>). Though we commonly categorize these insecticides according to their chemical classification (e.g., organophosphate, carbamate, pyrethroid, etc.), <u>each should be considered and used for their specific attributes</u>. Chemical toxicity is

often the most emphasized characteristic, but toxicity is only a part of efficacy when targeting a specific pest.

Most modern insecticides have relatively short residual activity. These materials are subject to fairly rapid breakdown by environmental factors such as water hydrolysis, sunlight, heat, and microbial degradation. The effects of some of these factors can be lessened by formulation and special application technology. However, these factors still play a major role in determining whether an application will reach its full effectiveness potential.

Because pesticides have limited residual activity periods, <u>uniformly distributing the proper rate</u> of an insecticide at a time when the pest is most vulnerable is extremely important. <u>The life cycle of the pest determines the vulnerable period</u>.

Objective. The objective of applying an insecticide (or other agent) to control soil-inhabiting insect pests is to **deliver the active agent <u>into the zone of pest activity and feeding</u>** (the **Target Zone**). Toxification is achieved by contact with and ingestion of organic matter (i.e., thatch) and soil particles with molecules of insecticide attached. Mortality occurs when the active ingredient is removed in the insect during the digestive process.

Distribution. For best distribution, <u>liquid materials</u> should be applied as coarse sprays. Though finer sprays of herbicides are used in weed control, insecticides are subject to drift and tend to volatilize more rapidly once the spray reaches the turf. A spray volume of 3.0 to 4.0 gallons per 1000 ft² has been commonly used in the lawn service industry. However, since golf courses usually use boom sprayers that apply at rates significantly lower (1.0 gallon), care must be taken to **accurately calibrate** the sprayer and **uniformly apply** the material. Spraying at such rates for some pests may require the application of irrigation before the spray dries on the grass blades to avoid photo-degradation and maximize potential effectiveness.

Shower-type nozzles deliver coarse sprays.

Large spray droplets have a better chance of reaching the thatch. Immediate irrigation can help move the application to soil-inhabiting targets.

Granular spreaders can effectively apply controls in an even distribution, but must be calibrated for each product. Granules must be swept from the sidewalks onto the turf.

Movement of control materials into the upper soil layer occurs more readily in turf with no thatch.

Turf with an inch or more of thatch can almost completely stop control materials from reaching the activity zone of a soil-inhabiting target. In such extreme cases, it is probably best to renovate and follow agronomic practices that keep thick thatch layers from developing.

When <u>granular formulations</u> are applied, **grass blades should be dry** at the time of application so the insecticide particles bounce off the blades and sift as deeply as possible into the turf canopy. This brings the concentrated particle closer to the target and also provides protection from breakdown by UV light and other factors. While the urgency to irrigate following granular application is not as immediate as for liquid application, it should be done as soon as possible. From 1/4- to 1/2-inch of water should be applied, but puddling should not occur.

Binding. All insecticides have some affinity to adsorb (bind) to organic matter. Since most turf has some thatch, this layer of living and dead material, consisting of 30%+ organic matter is one of the major barriers to the insecticide reaching a target such as grubs.

Analysis of pesticide residues found in thatch immediately after an application and irrigation indicate that 95.0 to 99.9% of the insecticide remains in the thatch. Therefore, **if thatch thickness exceeds 1/2 inch, insecticide applications to control soil-inhabiting pests will often be ineffective**.

Residues of insecticides (in ppm) 24 hours after application (2 Sept 86) at different levels of a turf profile after being sprayed (at labeled grub rate) and immediately watered in with 1/2-inch irrigation and allowed to dry. (H.D. Niemczyk, 1989, unpublished)

Insecticide	Thatch	0.0 to 1.0-inch	1.1 to 2.0-inch	4.1 to 5.0-inch	LD70*
Bendiocarb	13.64	0.37	0.22	0.09	1.92
Carbaryl	20.09	0.05	0.01	0.02	2.26
Chlorpyrifos	20.11	0.06	0.01	<0.01	13.40
Diazinon	47.62	0.08	0.02	0.05	0.40
Ethoprop	30.38	0.58	0.17	0.29	7.45
Fonofos	11.02	0.07	0.02	0.09	4.15
Isazofos	9.66	0.01	<0.01	<0.01	1.06
Isofenphos	19.29	0.42	0.22	0.10	2.30
Trichlorfon	16.57	0.78	<0.01	0.03	2.86

*Lethal dose to kill 70% of a Japanese beetle grub population, in soil, as determined in laboratory dose-mortality tests.
Thatch = 0.75 inch.

As can be seen in the above table, 95 to 99% of the insecticide residues are in the 0.75-inch of thatch. Residues in the first inch of soil under the thatch are generally too low to have much lethal effect on Japanese beetle grubs (see the LD70 column). We speculate that most of the grubs that are killed are feeding in the soil-thatch interface where sufficient residues of the insecticide are present. The obvious question is, do the residues move down over time? See the following table.

Retention of residues of nine insecticides in thatch following application to a golf course fairway (at labeled grub rate) and immediately watered in with 1/2-inch irrigation (2 Sept 86). (from H.D. Niemczyk, 1986)

Insecticide	Water Solubility*	% Recoverable Residue in Thatch 1d	3d	1w	2w	4w
Bendiocarb	40	95	99	99	89	88
Carbaryl	50	99+	99+	99	99	96
Chlorpyrifos	2	99+	99+	99	99	99
Diazinon	40	99+	99+	99	99	99
Ethoprop	750	97	99+	99	99	94
Fonofos	13	99+	99+	99+	99+	99+
Isazofos	150	99+	99	99	99	94
Isofenphos	20	97	96	98	99	96
Trichlorfon	154,000	94	78	85	49	ND

* Water solubility of technical material in parts per million (ppm).
ND = not detectable   Thatch: 1.0 inch thick

From the above table, it is evident that, even after four weeks, most of the insecticide residues remained in the thatch, even with moderately soluble insecticides such as ethoprop and isazofos. The highly soluble insecticide, trichlorfon, did move down somewhat, but this insecticide degrades rapidly and residue levels in the thatch or underlying soil after one week were very low.

Tank pH. Several insecticides (e.g., trichlorfon and carbaryl and other carbamates) break down rather rapidly at pH of 8 or higher. The pH of the water used to prepared tank mixes varies with the time of year and location, especially if it is drawn from local ponds or rivers.

Though it is advisable to have the water supply tested periodically for pH and buffering capacity, **it is more important to check the tank mix after a pesticide has been added**. Most pesticide formulations contain buffering chemicals that adjust the pH to a proper level. A simple pH testing meter or paper (litmus paper) can be used to check the tank mix **AFTER THE PESTICIDE IS ADDED**. This pH reading will show if the tank mix has changed after mixing.

Products are available to adjust pH levels if necessary. For most insecticides a low pH is not a problem, but pH of 8 or higher is.

Running the irrigation system on the syringe cycle immediately after a spray application and before using the regular irrigation cycle helps wash the treatment off the grass leaf blades and maximizes the potential for control of soil-inhabiting insects.

Dissipation of insecticides (ppm) in aqueous media prepared from thatch from plots overtreated 2 Sept 86 with one of the following insecticides established on a golf course fairway receiving five previous annual applications of isofenphos (=Oftanol). (from H.D. Niemczyk, 1986)						
Insecticide	**0DAT**	**1DAT**	**3DAT**	**6DAT**	**8DAT**	**13DAT**
isofenphos	8.5	7.7	0.5	0.7		
carbaryl	10.1	0.2	0.2			
diazinon	8.4	9.3	2.0	0.4		
isazofos	9.5	9.5	8.0	0.4	0.1	ND
trichlorfon	10.0	9.8	7.8	9.9	8.4	9.0
ethoprop	9.3	9.6	9.3	9.5	9.2	9.1
bendiocarb	9.3	9.7	9.2	9.7	9.0	9.1
fonofos	9.3	8.7	9.2	9.0	8.8	8.9

DAT = days after treatment   ND = not detectable
Thatch: 1.0 inch thick
Other studies showed that turf with no previous history of isofenphos treatments had active residues for 100+ days.

Watering-In.

While the performance of all insecticides is reduced by thatch, timely use of irrigation can help achieve maximum control from treatments. Liquid insecticides should not be applied to dry thatch or soil when control of a soil-inhabiting pest is the objective. Under such circumstances, the spray is likely to evaporate or become tightly bound before penetrating the surface. If not moist from a previous rain or irrigation, irrigate the day before treatment is scheduled.

For both liquid as well as dry applications, an additional 1/4- to 1/2-inch of water applied immediately after treatment maximizes control potential. Posttreatment irrigation or rain is essential to moving the insecticides off the grass blades or granules and to the TARGET. Delays in watering-in the treatment can significantly reduce effectiveness.

Most golf courses would have difficulty irrigating the entire course after a liquid application. Use the syringe cycle to briefly wash the spray droplets from the grass blades into the thatch as each fairway is treated, then run the normal irrigation cycle to apply at least 1/4-inch of water as soon as possible.

Accelerated Microbial Degradation.

Most modern insecticides registered for turf are broken down by pH and water (hydrolysis), sunlight or by the action of microbes. This microbial degradation is performed by various bacteria, fungi and other organisms found in soils and thatch and should be considered to be a natural, beneficial action. However, some of these microbes can adapt to pesticide residues and develop the capability to break down (consume them) much faster than normal. This phenomenon is called *accelerated* or *enhanced microbial degradation*.

While this phenomenon was well known in the scientific community and in certain agricultural settings for its effect on herbicides, fungicides and insecticides, it was not (and still is not) commonly discussed in the turf industry until the insecticide, isofenphos, was used. This organophosphate insecticide was remarkable because it originally provided "season-long" control of white grubs. Efficacy was demonstrated to last for over 300 days in some studies. After two to three successful applications, isofenphos failures were reported. Resistance to the insecticide was not the problem. Microbes, mainly bacteria, had adapted to the residues and simply degraded (consumed) the insecticide before it could reach the target. Thus, the phenomenon of *enhanced* or *accelerated degradation* in turf was confirmed.

Since then, other turf insecticides and one nematicide have been shown to be susceptible to accelerated degradation. The residual life of some insecticides, like trichlorfon and bendiocarb are too short for this phenomenon to develop. Yet chlorpyrifos, which also has a long residual, has shown no evidence of being affected even after many years of use in turf and agriculture. **Each chemical appears to have its own attributes that may allow it to be more or less susceptible to enhanced microbial degradation**. We have much to learn about this phenomenon and the extent to which it affects the efficacy of fungicides, herbicides and insecticides used in turfgrass management.

Evaluate Efficacy.

A mistake often made when using insecticides is to ASSUME the application worked. Turfgrass managers usually expect to see dead grubs a few days after treatment or dead mole crickets on the turf surface the next morning.

Many factors determine the time necessary for an insecticide to reach the zone of insect activity in a concentration sufficient to cause insect mortality. Furthermore, because of variations in the nature and depth of thatch and upper soil, the concentration reaching this zone at any one point is not uniform. With time and additional rain or irrigation, this concentration reaches a maximum and a degree of mortality (control) is achieved. The length of time a lethal concentration remains in the zone of activity (residual) can vary from a few days to a few months. Check for the level of control one and two weeks after application.

No insecticide registered for control of soil-inhabiting insects will consistently provide 100% control. Most will produce a 70 to 90% reduction in the population. Therefore, if an unacceptable number of pests remain after two weeks, control was most likely not achieved because of one reason - **the pesticide did not reach the zone of pest activity or target pest**. Thatch, delays in watering-in and poor application technique are often common causes of failure. Except for the case of isofenphos, where accelerated degradation is known, the extent to which this phenomenon reduces the effectiveness of other insecticides is unknown.

New Insecticide Classes. The "traditional" organophosphate and carbamate insecticides tend to work best when soil-inhabiting insects are in their early stages and actively feeding close to the soil-thatch interface. The typical first instar white grub is about 1/60th the weight of a mature third instar and requires considerably less insecticide to cause mortality than one in later stages.

Traditional insecticides usually produce some insect mortality in a few days. Though most insecticides have some contact activity, **ingestion is the primary mode of toxicity**. Once ingested, the insect stops feeding, loses body fluids, undergoes rapid motion and soon dies.

While the organophosphate and carbamate insecticides have had a history of use and effectiveness, current concerns regarding their impact on the environment and human health indicates most may eventually be withdrawn or lose their registration for use on turfgrass.

A new class of insecticides, **neonicotinyls**, contains two subclasses, **chloronicotinyls** and **thianicotinyls**. These insecticides block nerve impulses at the postsynaptic site causing a complete <u>disruption of normal behaviors</u> such as feeding and movement. Starvation is usually the cause of death. The residual half-life of these new classes in soil ranges to 120 days and most are taken up by plant roots and concentrated in leaves and stems. These characteristics are conducive to their usefulness in controlling a **broad spectrum** of soil, thatch, leaf and stem inhabiting insect pests of turfgrasses. Though contact toxicity occurs, **<u>ingestion appears to be the primary mode by which insects acquire the insecticide</u>**.

Imidachloprid, a chloronicotinyl registered in 1994, was the first in this class to be labeled for control of turfgrass insect pests. This insecticide is residually active against early stage insects for much of its 120-day half life in soil. This characteristic is most useful in control programs designed to prevent damage from soil inhabiting insects such as grubs. Single applications made in May have effectively controlled the major species of grubs and bluegrass billbug through September. A <u>broad spectrum</u> of thatch, stem and leaf-inhabiting insects are also controlled or suppressed during this time.

Thiamethoxam, a thianicotinyl registered in 2000, **is the first in this class to be registered**. It also appears to have a 100+ day half life in soil and controls or suppresses a broad range of thatch, stem and leaf-inhabiting insects.

Whether "traditional" or new classes of insecticides are used, the earliest feeding stage of a soil-inhabiting target is usually the easiest to control. These masked chafer stages (left to right) - egg, newly hatched first instar grub, mature second and third instar grubs, pupa and adult - range from easy to control to nearly impossible. The egg and pupal stages do not feed and therefore do not ingest insecticides. The first instar grub is about 1/20th the body weight of the second instar and 1/60th the weight of the third instar.

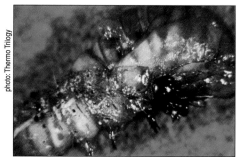

Malformed cutworm pupa caused after larva was exposed to the natural, botanical IGR insecticide, azadiractin.

Insect growth regulator (IGR) insecticides have been know for many years, but none have been used to control soil inhabiting insects in turf. <u>This class of insecticides act by mimicking insect hormones that control molting and growth</u>. These hormones are normally present only when the insect is physiologically ready to molt. Once the hormone mimic (the insecticide) is taken in, the insect stops its normal feeding behavior and may <u>molt before it is ready</u> to or the new exoskeleton doesn't develop properly. In either case, the insect dies. IGRs used against social insects, such as ants, often interfere with the reproduction of the queen(s). Generally, IGRs are most effective when the target insects are in their early stages of development. Later stages stop feeding but death may not occur for some time after treatment. <u>The spectrum of target insect pests affected by IGRs is apparently narrower than that of the chloronicotinyls and neonicotinyls</u>.

The IGR **halofenozide** was registered and labeled for control of insect pests of turfgrass in 1997 and is effective against some species of grubs. Preventive applications in June provide control of Japanese beetle and masked chafer grubs through September, but control of European chafer and bluegrass billbug larvae may be limited. Curative applications are reported to be effective against second instar Japanese beetle and black turfgrass ataenius grubs.

Unlike carbamate or organophosphate insecticides, imidachloprid, thiamethoxam, and halofenozide have **low impact on beneficial organisms** that live in turf. <u>In fact, imidachloprid inhibits the defensive behavior of grubs, making them more susceptible to natural control agents</u> such as bacteria, fungi, and nematodes that commonly inhabit the turfgrass environment.

Biological Control

Soil-inhabiting insect pests have many predators, parasites and diseases that attack them. Many of these are naturally occurring organisms but most rarely act rapidly enough alone to adequately control pests. Because of this, biological agents are usually reared and released into the environment having the target pest. This process is called <u>augmentation</u>. In a few rare instances, biological control agents not normally present have been <u>introduced</u> into the turf habitat. These foreign introductions are most commonly used to control introduced pests.

Predators and Parasites. Though naturally occurring **predators** and **parasites** are occasionally effective in suppressing turf insect pests, they usually have unreliable and unpredictable efficacy. In some cases, predators and parasites, such as spiders and wasps, cause more concern than the target pest.

Fire ants are considered by some to be beneficial predators since they commonly attack pest insects such as this mole cricket, but most would prefer not to deal with fire ants!

Illustration of Scolia dubia, a common parasite of the green June beetle grub. Most people are afraid of this wasp.

Ground beetles are common predators in turfgrass habitats. They have been known to feed on eggs, grubs and mole crickets. Again, most people consider them "unattractive," nuisance pests!

The wasp, *Scolia dubia*, a parasite of the green June beetle, has been be effective in lowering grub populations below damaging levels. Unfortunately, these large, often hairy wasps can cause great alarm in people afraid of being stung. Actually, these wasps rarely sting humans, and then only when severely provoked. Other parasitic wasps, *Tiphia* spp., parasitize masked chafers, May-June beetles and Japanese beetles, but most of these wasps rarely control their hosts on a regular basis.

Pathogens.
Insect **pathogens** (diseases) have been the most promising and effective biological control agents. Bacteria and fungi are often easily reared or produced in artificial media. Many of these microbes also produce resistant spores that may be formulated and distributed. In the past, only a few pathogens were commercially available but recent production and formulation techniques are making more available.

Milky Disease of White Grubs. Many species of white grubs are susceptible to infection by the bacteria, *Paenibacillus popilliae* and *P. lentimorbus*, that causes milky disease. These bacteria have different strains that infect different species of white grubs. The commercial products currently available contain only the strain infective in Japanese beetle grubs.

Infection takes place when resting spores of the bacterium are ingested by the white grub along with food and soil particles. The spore germinates inside the grub, producing bacteria which multiply and produce millions of new spores that eventually fill the entire body cavity of the insect. Infected grubs may live for months but eventually die. When dead grubs rupture, new spores are released into the soil. Yearly infection and subsequent spread of spores by dying grubs is the primary way further distribution occurs. The spores remain viable in the soil for many years.

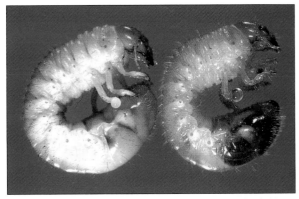

Milky grub (left) and normal (right). Note the milky drop of blood from the leg of the infected grub and clear drop from the normal grub.

Commercial preparations of milky disease are made by collecting thousands of living Japanese beetle grubs, infecting them in the laboratory, waiting until the bacteria reproduce, and then grinding and blending the grubs with talcum powder. The resultant dust can be applied to the turf any time the ground is not frozen. A common method of distribution is to apply one teaspoonful (about 200 million spores) of dust in spots at intervals of four feet in rows four feet apart. This takes about 12 pounds of dust per acre. This strain is not effective against other species of grubs.

Application of Japanese beetle milky disease powder.

The spore powder application usually takes 3 to 5 years to reach maximum infectivity. Unfortunately, recent field evaluations of this pathogen have indicated that it may be a relatively "weak" pathogen, usually infecting only 20 to 50% of a grub population. Evaluation in New England states often yield 40 to 50% infection while studies in Ohio and Kentucky show infectivity rarely reached 25 to 30%. Soil and weather conditions in different regions of the country may greatly affect the level of control realized by this bacterial disease.

Milky Disease of BTA. Black turfgrass ataenius (BTA) grubs are susceptible to infection by a specific strain of *Paenibacillus popilliae*. Occasionally, significant numbers of BTA larvae are infected with this naturally occurring strain. Infected larvae are opaque white and, though still alive, are not as active as healthy grubs and have stopped feeding. The presence of infected grubs should be viewed as "good news" because dead larvae release millions of spores that remain to infect future generations of BTA.

Black turfgrass ataenius grubs infected with milky disease often "survive" insecticide treatments.

The BTA-infective milky disease has caused some confusion. Examination of turf one to two weeks after application of an insecticide for control of BTA larvae sometimes reveal considerable numbers of milky grubs. Apparently, the milky larvae survive the insecticide while the healthy grubs are killed. One might conclude that the insecticide failed and another application is warranted. In fact, this result is ideal since healthy grubs are killed and the surviving milky disease infected grubs will die of the disease and add to the spore residue in the soil. Our theory is that **infected grubs stop feeding and therefore survive because they do not ingest the insecticide**.

BT for White Grubs. The bacterium, *Bacillus thruingiensis* (commonly called "BT"), is a common bacterium found in soils. When first discovered, some strains contained toxin granules that seemed to affect the gut lining of certain insects, mainly leaf feeding caterpillars such as sod webworms and armyworms. Since discovery of this original group of strains, over 10,000 different strains have been characterized.

The white grub active strain is *BT* variety *japonensis*, strain 'buibui.' Initial field tests have shown promising efficacy (70 to 90% control) against Japanese beetle and masked chafer grubs. Other grub species may be controlled, but to a lesser extent. At present, no commercial preparations of this strain are available.

BTs are used differently than milky diseases. BT bacteria are easily produced in large fermentation tanks, **in vitro**. Spores of BT are fairly susceptible to degradation from sunlight and other microbes may kill the spores. Therefore, this BT is applied using "inundative augmentation." In other words, large numbers of spores are applied to the area where white grubs are active, similar to broadcast application of an insecticide. The infection is rapid and maximum control is achieved in a few days to two weeks. Though some BT spores may survive until the next season, most simply decompose. Annual application is needed whenever the grub populations reappear.

Amber or Honey Disease of White Grubs. This disease is caused by the bacterium, *Serratia entomophila*. Infected grubs stop feeding and their body fluids become a honey-amber color. Affected insects become flaccid in a few weeks and soon decay. Though *Serratia* is found around the world (including the United States), the only

Grub infected with Serratia (left) and normal.

commercial product is available in New Zealand where it is successfully used for management of the grass grub in sheep pastures.

White Fungus Disease. The fungus, *Beauveria bassiana*, is often called the white fungus disease of insects because it covers infected insects with a snow white fungal mass. Infected insects become sluggish and eventually stop all activity. Within a few days or weeks the fungus that has been growing within the insect body sporulates by forming a dense cottony mass over the insect exterior. Many strains have been identified and almost all turfgrass-infesting insects are susceptible. Chinch bugs and billbugs are commonly infected and their populations severely reduced by this fungus.

White grubs infected by Beauveria, like the Japanese beetle grub above, often become unrecognizable masses of fungal mycelia.

Beauveria outbreaks often occur in periods of rainy, cool weather. Though white grubs and mole crickets are susceptible, their populations are rarely controlled adequately. A recent product, Naturalis-T™, has been marketed for management of mole crickets and chinch bugs.

Apparently, the key to obtaining maximum efficacy from *Beauveria* application is to keep the turf thatch and soil moist for a week to 10 days after application. Naturally occurring ***Beauveria is also most effective where the turf is kept moist***. Application of fungicides destroys this fungus.

Bluegrass billbug adult infected with Beauveria.

Green Fungus of Insects. *Metarhizium anisophiae* forms an olive-green spore coating on infected insects. Infected insects become sluggish and stop all activity. The fungus grows within the insect body and soon coats the exterior surface with white mycelia. These mycelia sporulate, producing the greenish coloration.

Though several strains are being developed by foreign and United States companies for management of white grubs, no commercial products are yet available.

Entomopathogenic Nematodes. These nematodes are specialized roundworms which carry a bacterium lethal only to insects. Juvenile nematodes usually enter an insect through the mouth, anus or breathing pores, though some species may be able to penetrate through the insect exoskeleton. Once inside, the nematode regurgitates a bacterium. The bacteria multiply, generating a toxin that kills the insect and prevents other bacteria from colonizing the cadaver. The nematodes feed on the bacteria, mature and reproduce inside the body of the dead insect before releasing a new generation to seek other hosts. These nematodes are not harmful to animals other than insects and **they can not enter plant tissues**.

Steinernema nematodes are commercially available under several trade names. *S. carpocapsae* is the most commonly produced species because of the ease of producing juveniles in large fermentation tanks. *S. carpocapsae* is most useful for control of cutworms, sod webworms, billbugs and fleas. However, the nematodes are very susceptible to desiccation, can not tolerate direct sunlight, and may be killed by certain insecticides or fungicides applied to turf.

S. feltiae and *S. glaseri* are also marketed for surface insect and grub control. Steinernematid nematodes, in general, have not performed well for control of grubs.

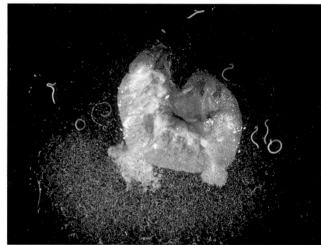

Japanese beetle grub infected with Steinernema carpocapsae. The larger nematodes are reproductive adults while the very tiny ones are new infective juveniles

Three green June beetle grubs recovered from their burrows after being infected with Metarhizium anisophiae. The green color is from the spores formed on the surface, the fungus growing inside the bodies is white.

S. riobravae and *S. scapterisci* are species registered for control of mole crickets and properly made applications have produced satisfactory control.

Heterorhabditis bacteriophora nematodes are commercially available from smaller suppliers. Recently, larger scale production of this nematode under the name of Cruiser™ has been accomplished. Heterorhabditid nematodes have generally been the best performing species for control of grubs.

Best efficacy has been achieved when the nematodes are applied in the late afternoon or evening (to avoid exposure to direct sun light), to moistened thatch and soil, followed with immediate watering-in.

Biological Controls and the Target Principle

The essence of the **Target Principle**, as described earlier in this chapter, is that materials directed at controlling damage from soil-inhabiting insects must reach their primary feeding and/or activity zone to be effective. Current methods of applying biological control agents for pests such as grubs, usually involve broadcast or spot applications to the turf surface followed by irrigation or rain to move the agent to the activity-feeding zone of the target insect. A major reason for the limited (at best) success of these agents is that they simply do not reach the **Target** in sufficient quantities to be effective. The physical condition and characteristics of the medium above the Target presents a major impediment to their downward mobility. Until methods and equipment are devised that deliver these agents directly into the **Target Zone**, without unacceptable damage to the turf surface, biological controls will continue to have limited success and acceptance.

Cultural Control

Quarantine. Japanese beetle quarantines are currently operated by the USDA-APHIS-PPQP and states involved with shipping materials out of infested areas into uninfested areas. Though this has not stopped the slow progression of Japanese beetles westward, it seems to have slowed the process. Nursery plant and sod producers shipping plant material with soil out of Japanese beetle infested areas must obtain an inspection and certification. Often airports and rail yards are under quarantine and transporters must treat their containers before shipping.

The European chafer is a serious pest of nursery stock. Using planting stock certified free of this and other root pests helps further reduce their spread. Other than allowing the soil to dry out during the time eggs are developing, no other cultural controls have much influence on this turf pest.

Thatch Management. Thatch is a mixed blessing. On the one hand, it is a major obstacle to delivering control materials to target insect pests like grubs. On the other hand, thatch significantly reduces the potential for ground water contamination by pesticides and serves as a reservoir for insecticides applied to control pests such as chinch bugs, billbugs and mole crickets that live in it.

Thick thatch removed from soil core sample.

Core aerification, top dressing, verticuting and reducing fertilizer use can help reduce the build up of thatch.

Commercial Japanese beetle pheromone trap.

Black turfgrass ataenius and *Aphodius* larval infestations usually occur in thatchy turf. Management practices that help reduce thatch and compaction may help in reducing the chance of infestation. Occasionally, when thick thatch exists and normal management practices are no longer effective, a complete dethatching or renovation (removal) may be necessary.

Habitat Modification. Eggs and young grubs are very susceptible to desiccation in dry soils. Therefore, omitting irrigation during the time eggs and first instar larvae are developing is detrimental to the insects. While this tactic is generally impractical for golf courses, it may have some application in other turfgrass situations. If natural rainfall occurs, this approach is nullified.

Trees or shrubs highly attractive to adult Japanese beetles near turf should not be planted, especially along golf course fairways and surrounding athletic fields. Trees and shrubs most attractive to adults include: grape, linden, Japanese and Norway maple, birch, pin oak, horse chestnut, Rose-of-Sharon, sycamore, ornamental apple, plum and cherry, rose, mountain ash, willow, elm, and Virginia creeper. Trees and shrubs rarely attacked include: red and silver maple, tuliptree, magnolias, red mulberry, forsythia, ash, privet, lilac, spruce, hydrangea, and taxus (yew).

Masked chafer adults are attracted to ***lights*** at night and grub damage is often common under or near street, athletic field, or other bright lights. Replacement with sodium vapor or yellow lights will reduce attractiveness.

Traps. Various traps have been developed to capture certain grub adults and mole crickets. Adult beetle traps use pheromones or chemical lures to attract beetles while mole cricket traps use sounds that mimic the call of males to attract females.

The commonly available Japanese beetle trap uses a beetle aggregation (floral) pheromone plus a sex lure to attract both sexes of beetles. Experimental tests with this trap indicate that the trap may attract beetles from 1/2 mile away, but can not collect all that are attracted to the area. When this trap is placed in the vicinity of susceptible ornamental plants, more damage can occur from feeding by attracted adults than if no trap is used. Japanese beetle traps are not recommended for control of grubs.

Mole cricket traps consist of a 4- to 6-foot diameter plastic wading pool with water over which is suspended an electric caller or tape recorder that produces the trill call of either the tawny or southern mole cricket. Mainly females are attracted, but males will also respond. The crickets fall into the water in the wading pool and drown within a day or two. The trap is more useful for monitoring mole cricket flights than as a control measure.

photo: B. Joyner

Mole cricket traps used to monitor adult flight activity. The boxes over each plastic wading pool are actually speakers that produce sounds similar to the chirps of the tawny and southern mole crickets. The pools are filled with water upon which the captured mole crickets float until counted and removed.

Linden trees, a favorite food plant of Japanese beetles, are severely skeletonized by the adults. Planting many such trees around high quality turf increases the probability of grub damage.

Resistant - Tolerant Turf. Turfgrass varieties with extensive root systems often have some tolerance to soil-inhabiting insects. Tall fescues can commonly tolerate annual grub populations in excess of 15 per square foot, while bluegrass-ryegrass blends may be damaged by eight to ten grubs per square foot. Among southern grasses, Cavalier® zoysiagrass is reported to have resistance to the tawny mole cricket.

Endophyte enhanced perennial ryegrasses and fescues have been shown to be quite resistant to leaf and stem attacking insects. However, the toxins produced by the endophyte, a fungal symbiont that lives between the cells of the leaf and leaf sheath, are not translocated to the root systems. Except for bluegrass billbug, soil-inhabiting insects such as grubs are apparently not affected by endophytic grasses.

Adults and young larvae of billbugs feed in and on the stems of grasses and therefore ingest endophyte toxins. There is ample evidence to show that endophyte enhanced perennial ryegrasses and fescues can significantly reduce populations of bluegrass billbugs. Overseeding or replacing the turf with blends of Kentucky bluegrass and endophyte enhanced grasses is an effective cultural approach for control of billbug. Generally, a turf stand with 30-40% endophytic plants is sufficient to control billbug damage.

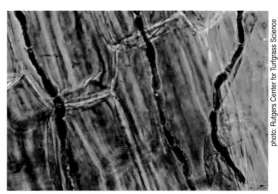

photo: Rutgers Center for Turfgrass Science

Microscopic view of perennial ryegrass leaf blade showing cells and threads of fungal endophyte mycelium (stained red) *growing between them.*

Notes

Notes

Chapter 5

Crown & Thatch-Inhabiting Pests

Sod Webworms

Over 30 species of sod webworms have been identified in North America. However, only half of these commonly occur in turfgrass areas and even fewer are considered pests. Pest species vary across the United States and can be grouped into those inhabiting cool-season grasses and those inhabiting warm-season grasses.

Species which prefer cool-season grasses are: bluegrass webworm, larger, western, striped, elegant and vagabond sod webworms, and the cranberry girdler. Some of these may also occur in the warm-season zones, but the imported tropical sod webworm is the principal pest of warm-season grasses.

Sod webworm moths are fairly easy to identify to species by using wing color patterns. Adults of most species rest on grass blades and in shrubs during the day and characteristically roll the forewings tubelike around the body. The head has a **snout-like projection extending forward**, thus, they are also called snout moths. Tropical sod webworm adults hold their forewings roof-like over the body. Usually two or three sod webworm species cause damage in any given area and species complexes vary across North America.

Typical sod webworm larva showing rows of shield-like spots on body. When uncovered in turf thatch, their green frass (fecal) pellets are a sign that they have recently eaten turf tissue.

The larvae may be light tan to light purple and often have a greenish cast due to the chlorophyll contents of their gut. Upon close inspection, all species have characteristic rectangular, shield-like spots in rows down the body. Larvae are quite difficult to identify to species and an expert should be consulted if larval identification is needed.

Adults of most species lay ribbed eggs by dropping them into the turf at night as they fly over it. The tropical sod webworm attaches flat scalelike eggs to blades of grass.

Diagnosis. Generally, cool-season sod webworm larvae construct tunnels in the soil and thatch and line them with silk. At night, the larvae follow the tunnels to the surface and feed on grass blades and stems just above the crown. The **severed stems die, leaving dead spots** and/or sparse and ragged appearing turf. However, **sod webworm larvae rarely kill turf**. Brushing or

Striped sod webworm (above) **and vagabond sod webworm** (right) **showing typical snout-like mouthparts and common resting position on grass blade.**

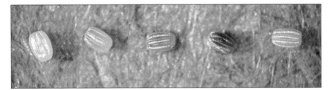

Larger sod webworm eggs are dropped into the turf at night. (left to right) **1, 3, 5, 10-day-old and hatched eggs.**

Overwintered sod webworm larvae feed on spring turf growth causing spots that do not "green up" normally. Similar damage occurs during summer dry periods.

Flocks of birds (especially starlings) that frequently return to a turf area usually means that sod webworms or other larvae are present. Further evidence of bird activity is probe holes left by birds searching for larvae. Close examination of the turf in such areas either reveals larvae, or the green pellets of excrement (frass) left by them.

raking away the dead stems exposes the shortened green stems left by the grazing larvae. The tropical sod webworm feeds along the tips and edges of grass blades, similar to armyworms. High populations can literally mow down turf.

An effective method of detecting infestations is to mix two tablespoons of liquid dishwashing detergent in two gallons of water (in our experience, Joy™ has not caused damage to the turf). Apply the solution uniformly over one square yard of turf using a sprinkling can (**= soap flush**). The soap solution irritates the larvae which come to the surface in 10 to 15 minutes. Early morning is the best time to sample because the larvae are close to the surface. **Larger larvae may surface first and the smallest last** (after 20 minutes). This method is least successful when the thatch and/or upper soil are dry. Preirrigation may help.

Life Cycle and Habits.
In cool-season turf areas, female moths simply drop their eggs as they fly over the turf at dusk and after dark. Eggs hatch in a week to 10 days and about six weeks are required for development from egg to adult in summer months. The most common species on northern turfgrass (bluegrass and larger sod webworms) have two generations each year and overwinter as larvae in silken webs (hibernacula) within the thatch or top inch of soil. The cranberry girdler has a single generation per year. The tropical sod webworm has multiple generations each year.

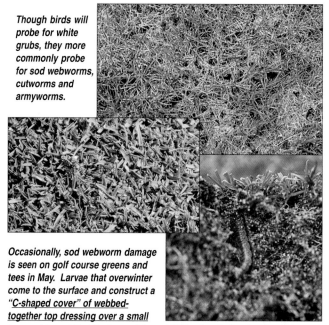

Though birds will probe for white grubs, they more commonly probe for sod webworms, cutworms and armyworms.

Occasionally, sod webworm damage is seen on golf course greens and tees in May. Larvae that overwinter come to the surface and construct a "C-shaped cover" of webbed-together top dressing over a small burrow (above left). The larvae (above right) feed on turf under the cover which gets larger as the larvae require more turf for food. This sand cover is just below the mowing level of the green.

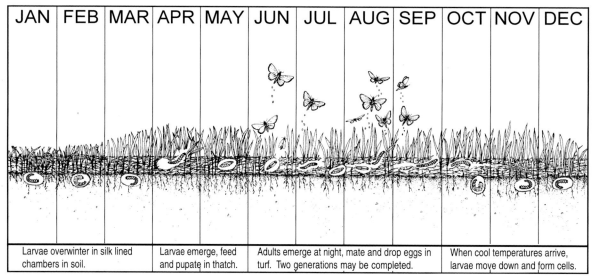

JAN	FEB	MAR	APR	MAY	JUN	JUL	AUG	SEP	OCT	NOV	DEC

Larvae overwinter in silk lined chambers in soil.	Larvae emerge, feed and pupate in thatch.	Adults emerge at night, mate and drop eggs in turf. Two generations may be completed.	When cool temperatures arrive, larvae move down and form cells.

Bluegrass Webworm Life Cycle in Ohio.

54

Bluegrass Webworm

The bluegrass webworm is found in the eastern half of North America. This species prefers Kentucky bluegrass but also feeds on ryegrass, fine and tall fescues, as well as weed grasses, such as crabgrass and orchardgrass.

Freshly hatched larvae are 3/64- to 5/64-inch (1 to 2 mm) long, have blackish-brown head capsules and a translucent yellowish body. After feeding, the body becomes greenish from food in the gut. As the larva molts and grows through a minimum of seven instars, the head capsule turns brownish-yellow tinged with green, the body color becomes straw yellow and the segmental spots are reddish-brown. Mature larvae are 3/8- to 1/2-inch (9 to 13 mm) long.

Adults have a wing span of 9/16 to 13/16 inch (15 to 21 mm). The snout and head are white above and gray to a light brown below. The most distinctive feature of the forewings is seven spots along the wing tip. A curved brownish-orange line runs across the wing just inside the tip, and the veins are usually distinctly lighter in color. The hind wings are lighter than the forewing and have a narrow brown line around the margin.

Bluegrass webworm adult.

Diagnosis & Damage.
Closely mowed turf shows symptoms readily. Single larval feeding can cause small depressed pock spots of brown grass. As the larvae develop, these spots may become several inches in diameter. In heavier infestations, dead patches coalesce to form irregular patterns of sparse turf. Poorly maintained turf may have a general ragged appearance with many scattered dead stems. Brushing away these stems reveals short green "stubs" of turf left by the grazing larvae. Birds feeding on the larvae often make probe holes in the dead turf. Infestations in golf course fairways and roughs as well as in home lawns are **easily misdiagnosed as summer dormancy and can be confused with billbug damage**.

Life Cycle and Habits:
Adults can be found during most summer months, but peaks in adult flights in June and August suggest two generations are normal in the Middle States. Larvae overwinter in silk-lined chambers (hibernacula) in the soil. In April, the larvae resume active feeding before pupating in late May and early June. Spots of turf that do not show normal spring green-

up and have bird probes often appear at this time. Pupation takes 5 to 15 days, depending on spring temperatures.

Adults emerge after dark and usually mate before sunrise. Males often die within a day, and females live only 5 to 7 days. Females begin laying eggs the night after mating by hovering two feet or less over the turf and dropping their eggs. Maximum egg laying occurs about an hour after sunset and may continue for two hours. A female may lay 200 eggs. The eggs can hatch in 5 to 6 days at 70°F, but take longer at lower temperatures.

Emerging larvae spin webbing along a leaf blade, eat the leaf surface tissues, and after a molt or two drop to the ground to form a tubelike silken tunnel with pieces of thatch attached. Maturing larvae feed at night near the tunnel opening and deposit green fecal pellets (frass) there. Larvae take about 40 to 45 days to mature during the summer. Second generation larvae maturing in the fall burrow deeper into the soil to overwinter in a silk-lined chamber.

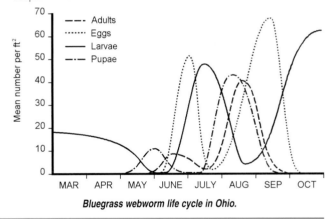

Bluegrass webworm life cycle in Ohio.

Larger Sod Webworm

The larger sod webworm is common in bluegrass and tall fescue growing regions of the northern half of North America. Larvae seem to prefer Kentucky bluegrass but may attack ryegrass, fine and tall fescues.

Maturing larvae go through a minimum of seven instars. The head capsule is brownish-yellow with darker markings and the general body color becomes yellowish with green food contents. Body spots are **distinctly chocolate-brown**. Mature larvae are 1-inch (24 to 28 mm) long.

Adults have a wing span of 7/8- to 1-3/8- inch (21 to 35 mm). The head, snout and thorax are straw colored and speckled with brown tipped scales. The forewing color is rather nondistinctive and

varies considerably from almost solid cream to grayish with light colored veins. The tip has silvery gray scales interrupted with white scales where the veins end. Usually, three small black spots are present at the tip.

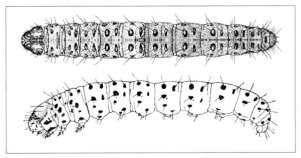

Diagram of larger sod webworm larva, top view and side, showing distinct dark brown spots.

Larger sod webworm adult

Damage and Diagnosis. Damage and damage
symptoms are the same as those previously described for the bluegrass webworm.

Life Cycle and Habits. This sod webworm
apparently has 2 to 3 generations per year with peak adult flights in mid-June, late July and mid-September. In the Pacific Northwest, this species is reported to have one generation. Larvae overwinter in silk-lined chambers beside turf roots. Feeding resumes for a short time in the spring and pupation occurs in May. Within 10 to 20 days, the adults emerge at night, with males appearing early after dark and females a few hours later. Mating continues until dawn.

Mated females begin laying eggs the night after mating and lay an average of 180 over 4 to 6 nights. Young larvae spin a loose net-like web in the fold of a grass blade and feed between two veins. After molting three times, larvae form webbing tunnels running along the ground, through the thatch and often into the ground. Larvae feed at night by chewing on grass blades, and usually leave green fecal pellets (frass) at the open end of the tunnel. When this tunnel becomes too fouled with frass, a new tunnel is constructed. Several of these tunnels may be constructed before the larva matures. At maturity the larva forms a cocoon. Larvae not maturing soon enough to emerge in the fall overwinter in cells constructed in the soil or thatch.

Western Lawn Moth

The western lawn moth is commonly found in the Rocky Mountain plateau west to the Pacific Coast and appears to be most common from Washington through California. This species prefers Kentucky bluegrass, perennial ryegrass, fine fescue, and bentgrass, but may shift to bermudagrass in southern climates.

The body of feeding larvae are greenish from food in the gut. The body spots are distinctly dark by the fourth or fifth instar. As the larva molts through a minimum of seven instars, the head capsule turns dark brown with darker mottling. Mature larvae are about 5/8-inch (15 mm) long.

Adults have a wing span of 11/16- to 15/16-inch (17 to 23 mm). The snout and head are white above and gray to buff below. The most distinctive feature of the buff colored forewings are the three elongate spots running along the middle. The forewing ends with a chevron mark, followed with a series of black dots. The hind wings are darker than the forewing and have a silvery cast.

Damage and Diagnosis. Damage and damage
symptoms are the same as those previously described for the bluegrass webworm.

Life Cycle and Habits. Adults occur during most
summer months, but peaks in adult flights indicate that 4 to 5 broods occur each summer. The second and third broods, in July and August, appear to cause the most significant damage. Larvae remaining by late fall overwinter in silk-lined chambers in soil or thick thatch. In the spring, these return to feed briefly and pupate in mid-May.

Western lawn moth adult. Note elongate white spots on wing.

Adults emerge after dark and mate during the first night. Males often die within a day or two, while females begin laying eggs the night after mating and live 5 to 7 days. Eggs are dropped into the turf canopy as females hover.

Eggs hatch in 5 to 10 days, depending on the temperature. After a molt or two, young larvae drop from turf leaves to the ground to form a tubelike silken tunnel. The tunnel often has piles of green fecal pellets near its opening. Larvae feed at dusk to dawn and take about 30 to 40 days to mature during the summer. A complete generation takes about 45 days, and 3 to 4 generations can occur over the summer. The second and third generations appear to be the largest. These larvae feed from June through early August.

Tropical Sod Webworm

The tropical sod webworm <u>occurs worldwide in the tropical zones</u>. Larvae feed on turfgrasses in Florida and the Gulf States where freezing temperatures are not reached. St. Augustinegrass, bermudagrass and centipedegrass are primary hosts, but zoysiagrass and bahiagrass may be damaged.

<u>Adults</u> are <u>very different</u> from regular sod webworm adults because the <u>wings are held roof-like over the body</u> instead of being rolled around the body. The body shape looks much like a <u>swept-wing jet</u>. Adults may be dingy brown to straw colored and have irregular darker marks. The wing span is about 13/16-inch (20 mm).

Typical tropical sod webworm adult (left) and other similar species (right) commonly occur together in southern turf. We believe that some work is needed to identify which species actually cause damage.

Damage and Diagnosis.
<u>Outbreaks seem to appear overnight</u> when the larger larvae begin to forage on turf blades above ground. Early symptoms are <u>ragged edges on the leaves followed by stripping of foliage similar to armyworm damage</u>.

Larger larvae feed at night and hide below the grass surface during the day. Parting the turf at the edge of the damage margin, reveals the resting larvae curled at the soil surface. The <u>larvae leave trails of silk</u> as they move from one grass blade to another. If dew is present, these <u>webs are easily seen</u> in the morning.

Life Cycle and Habits.
The tropical sod webworm has <u>continuous generations</u> during the year and slows its cycle during cooler temperatures. Outbreaks occur most commonly during the hotter summer months following rainy periods. <u>Females lay eggs on grass blades or overhanging vegetation of ornamentals during the early night</u>. Young larvae spin webbing in the V-shaped depression of a blade of grass, feed on the leaf surface, and move to new blades of grass as needed. By the third instar, the larvae drop into the turf canopy and spin loose webbing between the grass stems. Larvae feed on the edges of the grass blades leaving a ragged edge that can be difficult to detect in healthy turf.

Tropical sod webworm damage to St. Augustinegrass first appears as a general thinning and close inspection reveals that the grass blades have chewed or notched margins. This damage is similar to armyworms. Severe infestation can strip grass blades completely.

In the last 10 to 12 days of development larvae eat entire leaves. Damage becomes obvious and radiates from where the eggs were deposited. At maturity, the larvae spin a loose bag of silk within the thatch and pupate for 7 to 14 days.

Adult moths emerge at night, mate in the evening, and <u>do not fly during the day unless disturbed</u>. Adults are strongly attracted to lights and are not active at low temperatures. This insect is very sensitive to low temperatures and does not overwinter where freezing temperatures regularly occur. Populations decline when soil temperatures to drop below 60°F.

Tropical sod webworm larva look similar to other sod webworms except that the characteristic spots are smaller and light colored.

Cranberry Girdler

Though technically a sod webworm, the cranberry girdler is more <u>subterranean in habit and rarely feeds on turf leaves</u>. **"Subterranean sod webworm" is another common name for this insect.**

This species is <u>found across North America in the cool-season and transition turfgrass zones</u>. Cool season grasses such as Kentucky bluegrass, bentgrass, and fine fescues are preferred.

<u>Adult moths</u> have a wingspan of 9/16- to 13/16-inch (15 to 20 mm), and are quite colorful for sod webworms. The forewings have a series of longitudinal brown and cream stripes following the wing veins, a silver chevron across the wing tip followed with ocher, a series of three black spots and a tip fringe of silver scales.

Cranberry girdler adult.

Damage and Diagnosis.
Larvae mainly feed on roots and stems and bore into grass crowns, causing small circular to large, irregular patches of dead turf. Heavy populations can cause general turf death **similar to grub damage.**

Larvae are not typical sod webworms because they lack distinct dark spots over the body. Cranberry girdler larvae are a dirty-white color, with a tan head capsule. Mature larvae are 5/8- to 3/4-inch (16 to 20 mm) long.

Life Cycle and Habits.
Adult cranberry girdlers are active fliers from late June to mid-August. Adults emerge in the evening or at night and the females release a sex attractant that calls in receptive males. Mated females begin laying eggs the following day. Unlike other sod webworms that lay eggs while flying, these females drop their eggs (up to 500) into the turf while resting on a plant. Eggs take 9 to 11 days to hatch.

Young larvae reside in the thatch or upper soil surface next to grass crowns. As the larvae grow, they bore into grass crowns and

Cranberry girdler larvae have sod webworm spots, but the spots are the same color as the body!

are known to feed on roots, stems and leaves. Webbed tunnels lined with fecal material are usually evident by late August and September. Mature and nearly mature larvae spin a tough silk case (hibernaculum) in soil or thatch in October and spend the winter in diapause (a condition similar to hibernation). Larvae that did not mature in the fall resume feeding in the spring. Fully mature larvae do not feed, but pupate with the rest of the population in May and early June. Depending on the temperature, pupae take 2 to 4 weeks to mature.

Burrowing Sod Webworms

Burrowing sod webworms are generally present east of the Rocky Mountains. Most activity is noticed in the transition and southern turfgrass zones. Very little is known about the members of this primitive group of moths. They are mainly tropical and subtropical in distribution with species living on bromeliads and orchids. Many species in North America apparently feed on the roots and stems of grasses. What little is known is based on observations of attacks on corn roots in no-till or minimum till corn fields. Host grass specificity has not been studied, but infestations have been recorded in Kentucky bluegrass, tall fescue and bermudagrass turf.

Adults are mottled brown or reddish-brown moths with wing spans of 1- to 1-3/8-inch (25 to 35 mm). Male snouts are very large, hairy and curved over the head to touch the thorax. Females have short snouts. Adults are often collected in light traps and are relatively easy to identify.

Burrowing sod webworm larvae build white, silken tubes to line their burrows during pupation. Birds commonly pull these to the turf surface.

Burrowing sod webworm male with characteristic snout extending back over the head and thorax.

Damage and Diagnosis.
Damage to turf occurs rarely and resembles that from other sod webworms or cutworm. Larvae build **white silken tubes over the surface of the thatch**. These are unsightly and are often pulled out during mowing or by birds. These tubes look like empty cigarette papers and may litter a lawn after the adults emerge. Larvae are tan to

pinkish-brown, have a velvet textured surface, no distinct spots and a characteristic, black platelike band across the thorax, behind the head.

Life Cycle and Habits.
Adults are active from mid-June through July, flying at dusk until an hour or two after dark. They move very swiftly and upon landing on the turf, quickly crawl or wriggle to the thatch surface. Eggs are dropped over the thatch surface. Upon hatching, young larvae establish themselves next to a plant stem and burrow vertically into the

Burrowing sod webworm larva in its burrow.

soil where they construct a silk-lined tunnel. Burrows may extend from 4- to 24-inches into the soil. Larvae extend from the burrows at night to feed on turf leaves.

The silk tubes may be extended above the soil-thatch and along grass stems. In very close cut turf, these tubes are occasionally built across the turf surface. Larvae spend the winter in their burrows but resume activity in the spring. At maturity, the tunnels may be enlarged from 5/32- to 1/4-inch in diameter. The pupa is located in the tunnel near the soil surface.

Mature burrowing sod webworm larvae have a velvety surface texture and a dark collar-like plate behind head.

Billbugs

Billbugs are gray, black to reddish-brown weevils, each of which has a ***characteristic, conspicuous snout*** with chewing mouthparts at the tip. Though over 60 species exist, only four are known to cause damage to turfgrasses in North America. The bluegrass billbug damages most cool-season grasses (especially Kentucky bluegrass and perennial ryegrass) from Washington to Utah to the East Coast of the United States. The hunting billbug occurs in the warm-season turf zone and damages bermudagrass and zoysiagrass. The larvae of these two species account for most of the damage done by billbugs.

Adult bluegrass billbugs like this are commonly seen "wandering" about on sidewalks, curbs, and driveways in September.

The Phoenician (=Phoenix) billbug is a pest of bermudagrass and zoysiagrass in southern California. The Denver (=Rocky Mountain) billbug is found in Kentucky bluegrass and perennial ryegrass from New Mexico through Colorado and into Utah and Washington.

Damage, Diagnosis & Monitoring. Billbug damage is similar to that caused by drought, disease and other insects. Careful examination (hands and knees) of damaged turf is the key to correct diagnosis.

Adult billbugs feed on grass stems and stolons, occasionally chewing transverse holes through them. Damage inflicted by this feeding is considered insignificant. Some species lay eggs in the feeding holes and others chew slits in stems and stolons and insert eggs in them.

The wandering nature of billbug adults is a good indicator of a potential problem in nearby turf. Careful observation for adults on sidewalks, driveways and along gutters during migration times should

Careful examination of turf, especially the "hands and knees" technique, is the best way to determine if drought, disease, billbugs or some other pest or agent is the cause of the damage. For billbugs, use the "tug test."

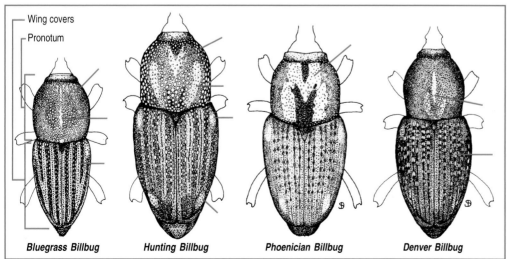

— Wing covers

— Pronotum

Bluegrass Billbug   **Hunting Billbug**   **Phoenician Billbug**   **Denver Billbug**

Billbug adults are easily identified by using the pattern and shape of smooth, raised areas on the pronotum, as well as the size of pits and furrows found on the wing covers. The bluegrass billbug has small to medium pits on the pronotum and often a median raised area; the wing covers have even rows of pits. The hunting billbug has a Y-shaped raised area surrounded by ( )- shaped marks; the wing covers each have two raised, smooth areas. The Phoenician billbug has a broad M-shaped area on the pronotum. The Denver billbug has a smooth median area on the pronotum and rows of paired pits on the wing covers.

provide warning of a possible infestation of larvae. When picked up or disturbed, adults fold their legs tight against the body and appear dead. Focusing the suns energy on their bodies using a hand lens soon revives them!

A good method of monitoring adult activity is to place small plastic cups (pit fall traps) inside holes made by using a 4-1/4-inch cup cutter. Traps can be placed along the turf margin, near flower beds so that they are out of the way. Adults falling into the traps can be easily counted by inspecting the traps 2 to 3 times a week.

Billbug larvae are white and legless soil, crown and root inhabitants, 3/8-inch long with a tan to brown head. They are fat with the tail end somewhat larger than the head end. Larvae feed on turfgrass roots, stolons, stems, and crowns, often cutting the stems off at the crown. If small patches of turf appear to be dying, the best way to determine if billbugs are responsible is by carefully examining the damaged turf. Turf damaged by larvae is easily pulled out by hand with the stems breaking off at the crown. A good indicator is the presence of fine, sawdust-like material (frass) left by larvae feeding in the crown or root zone. Look for evidence of larval feeding in the crown, at the base of the stem and use a knife to probe among the roots for larvae.

Life Cycles and Habits.
Except for the hunting billbug, all billbugs appear to have a single generation per year. The hunting billbug may breed continuously in southern states. Most billbugs spend the winter as adults, though significant numbers of hunting and Denver billbugs overwinter as larvae. For most billbugs, a spring adult feeding period is followed by egg laying, and larvae developing over the summer.

The "tug test" is an excellent method to confirm that billbugs have been feeding in grass stems and crowns.

After you have pulled up the dead grass stems, inspect the broken ends for evidence of sawdust-like frass around the broken ends - a sure sign of billbugs. No other insect causes this.

Bluegrass billbug larva at crown. Note sawdust-like frass trail down the stem to crown area.

Bluegrass Billbug

The bluegrass billbug is most common in northern North America, may be found in Southern States and is a common pest of lawns and golf courses. Kentucky bluegrass is the preferred host, but this pest has been known to infest perennial ryegrass, fine fescue and tall fescue.

Adults are 5/16-inch (7 to 8 mm) long with a black body. When freshly emerged, the adults are reddish brown. Mature adults are first covered with gray waxy scales which when worn off by movement through the turf, leave the insect generally blackish gray. The pronotum is covered with small uniform punctures and the wing covers have distinct longitudinal furrows with pits. If in doubt, species determination should be confirmed by a specialist as several species of billbugs may be present in turf.

The larvae are typical of weevils, having a robust, fat, white body with no discernible legs, and brown head capsules.

Bluegrass billbug life stages *(left to right): **small larva, mature larva, early pupa, mature pupa, callow (immature) adult, matue adult.***

Damage and Diagnosis.
On golf courses, light infestations in well kept turf result in small dead spots which **closely resemble the turf disease dollar spot**. On lawns, such **spots may also be misdiagnosed as those caused by sod webworm** because adult moths are sometimes flushed from the turf. A pull (tug) on the grass stems in such spots readily distinguishes billbug damage from other causes. Stems damaged by billbugs break off easily at the crown and show evidence of tunneling and whitish fecal material (frass) left by the larvae. Further examination of the crown and roots usually finds the larvae.

Larval infestations usually occur in late June into August when turf may be dormant from lack of moisture. These conditions mask larval infestations and are often attributed to turf dormancy only.

Symptoms of bluegrass billbug damage to lawns, like this in late June and July are very similar to those caused by drought.

Infested, turf stems are easily broken off and brushed away with a swipe of the hand. If only dormant, this does not happen.

Moderate or chronic infestations result in a thinner stand of turf each year, especially in Kentucky bluegrass. Heavy infestations result in complete destruction of the turf in August.

Overwintered bluegrass billbug adults wander about during May and June and again as new adults in September and October. Surveys of driveways, sidewalks, and curbs is one way of identifying potential infestation sites.

Life Cycle and Habits.

Over most of its range, this pest overwinters in the adult stage. Adults have been found overwintering in thatch, cracks and crevices in the soil, worm holes, next to buildings, in hedges or ground covers, and in leaf litter near turf. Hibernating adults become active in late April to mid-May when the soil surface temperatures rise above 65°F, and wander about at this time. After feeding for a short period of time, the female inserts single eggs into holes chewed in grass stems just above the crown. Females have been known to lay over 200 eggs, usually 2 to 5 per day. Females may continue egg laying into mid-June.

Eggs hatch in six days, depending on the temperature, and the larvae tunnel within the stems. If a stem is hollowed out while the larva is small, the larva may drop out and bore into another stem. Eventually the larva begins feeding on grass crowns and later moves to the roots and rhizomes. This is the point at which visible damage is noticed. **Damage is most severe when the turf is drought dormant.**

After 35 to 55 days, the larva moves 3 to 4 inches deep into the soil and pupates. The pupa gradually darkens and a reddish-brown, tineral (new) adult emerges in 8 to 10 days. New adults appear to be most common in August through September and they seek out suitable sites for overwintering in October. Adults have been observed trying to fly but flight has not been observed.

Apparently, a partial second generation has been observed. In October, 1984, H. D. Niemczyk observed numerous tineral adults emerging from sod production fields near Cincinnati, Ohio.

Bluegrass billbug egg inserted in stem.

Bluegrass billbug larvae feed at the crown, severing stems that die and cause brown spots in the turf.

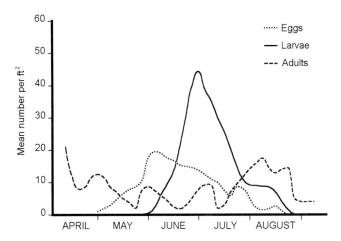

Seasonal occurrence of bluegrass billbug life stages in Ohio.

Hunting Billbug

Several subspecies of the hunting billbug occur across North America. The most common, *Sphenophorus venatus vestitus*, is found from Maryland across to Kansas and south. This pest was transported with sod to the Middle East, Southeast Asia and Hawaii.

Larvae attack zoysiagrass, bermudagrass, bahiagrass, centipedegrass, St. Augustinegrass, and it has been recovered from cool-season turf along the upper Atlantic Coast states.

Adults are generally larger and more robust than the bluegrass billbug. Adults range from 1/4- to 7/16-inch (6 to 11 mm) long, and often have a coating of soil adhering to the body surface. Clean specimens have numerous visible punctures on the pronotum with a **distinct Y-shaped, smooth, raised area behind the head**. This area is enclosed by a **shiny parenthesis-like mark on each side**. The adults commonly feign death when disturbed.

Hunting billbug adult. Note Y-shaped raised area on pronotum surrounded by ( )-shaped raised areas.

Larvae have the typical weevil form, are legless and 1/4- to 3/8-inch (7-10 mm) long when mature.

Damage and Diagnosis.

Zoysiagrass and improved hybrid bermudagrasses are damaged most severely. Infested turf shows patches of brown, dying grass. Damage is most severe during extended dry periods but can appear in irrigated situations, since damage to roots prevents moisture uptake. Adults and small larvae feed on stolons, crowns and leaf buds. Later stage larvae may damage roots to the degree that the sod breaks into pieces when lifted. The **pres-**

ence of sawdust-like fecal material inside hollowed out stems and stolons and around the root system, are sure indications of billbug activity. This billbug more commonly causes 6- to 12-inch diameter damaged areas that turn brown. When roughed up with your hand, stolons break up into short pieces with each piece showing evidence of being chewed off at the ends.

As with the bluegrass billbug, the hunting billbug adults are often seen on sidewalks, driveways and curbs on warm spring afternoons and in the fall. Surveys at this time are useful for anticipating potential infestations in nearby turf.

Life Cycle and Habits.
Little is known about the biology of this pest. In its northern range, this pest overwinters as a dormant adult in soil. In southern states, adults may be found walking and feeding all year whenever temperatures are high enough for activity. Generally, most eggs are inserted into grass stems or leaf sheaths in the spring when zoysia and bermudagrasses are well out of winter dormancy. Eggs take 3 to 10 days to hatch and the young larvae mine down the inner surface of the leaf sheaths and bore into stems. As the larvae increase in size, they feed on crowns and when mature, on roots and stolons. Mature larvae pupate after 3 to 5 weeks of feeding. The pupae take 3 to 7 days to mature.

Because of the extended period of oviposition, <u>larvae may be found from May into October</u>. In Florida and south Texas, larvae

Characteristic summer hunting billbug damage to bermudagrass lawn often looks like "doggie spots" (urine damage).

have been found overwintering. However, in the northern zoysia growing regions most damage occurs in August. Damage in Gulf States usually appears when the southern grasses begin to slow their growth in the fall. **Damage is also visible in the spring where billbug larvae overwintered, and is often mistaken for spring dead spot or delayed spring greenup.** Droughty conditions aggravate damage and masks symptoms.

Denver Billbug

This billbug is also known as the <u>Rocky Mountain billbug</u>. As its names suggest, this pest is mainly western in its distribution. It has been found from Nebraska to New Mexico and northwest to Oregon and Washington. In the northern states, it may be found with the bluegrass billbug.

<u>Larvae</u> are known to damage cool-season grasses, mainly Kentucky bluegrass and perennial ryegrass, and are typical in form to other billbugs.

<u>Adults</u> are larger and more robust than the bluegrass billbug, range from 5/16- to 3/8-inch (9 to 11 mm) long, and are usually **shiny black**, like patent leather. The small, sparse punctures on the pronotum and the **paired, heart-shaped punctures** on the wing covers are filled with white hairs.

Denver billbug adult.

Life Cycle and Habits.
Little is known about the biology and seasonal history of this pest. It appears to be active during much the same time as the bluegrass billbug. However, Denver billbug adults may lay eggs over a longer period of time in the late spring and summer and large **larvae commonly overwinter**. By late May most of the overwintered larvae have

pupated and the population is predominantly adults. Like most billbugs, eggs are laid in grass stems, usually a seed head stem. Early instar larvae burrow down the stem to the crown and later instars feed in the soil-thatch interface, much like white grubs. When mature, larvae burrow deeper into the soil and pupate.

Annual Bluegrass Weevil

The annual bluegrass weevil, also referred to as the "**Hyperodes weevil**" is an important pest of annual bluegrass, *Poa annua*, on golf courses and tennis courts in Northeastern States. The scientific name for the genus, *Hyperodes*, is in the process of change to *Listronotus*. The current preferred common name, "annual bluegrass weevil," is appropriate since *P. annua* is the only known host of this pest.

Turf damage has been localized in Connecticut, Long Island, Massachusetts, New York, Delaware, Pennsylvania, West Virginia and eastern Ohio.

Adults are small, 1/8- to 3/16-inch (3.5 to 4.0 mm) long, black weevils with wing covers coated with fine yellowish hairs, yellowish scales and scattered spots of grayish-white scales. Newly emerged adults (callow adults) are orange-brown in color and require several days before becoming fully black-pigmented.

The larvae are crescent-shaped, legless and have a creamy-white body. The head capsule is light brown in young larvae but becomes darker in older larvae which grow to about 3/16-inch (4.5 mm) when mature.

Annual bluegrass weevil damage in fairway.

Extensive annual bluegrass weevil damage to annual bluegrass surrounding green.

photo: H. Tashiro

photo: H. Tashiro

Adults feed on Poa annua leaves, making characteristic notches.

Larvae make U-shaped notches in the stem bases

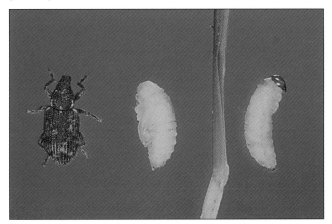

*Annual bluegrass weevil stages (left to right): **adult, pupa, Poa annua blade for reference, and prepupa.***

Damage and Diagnosis.
Adults emerge from overwintering sites, move to turf areas and **feed by cutting characteristic notches in the edges of grass blades** or holes in stems near the leaf bases. Examining the tips of grass blades at night with a flashlight is useful for locating the feeding adults.

Larvae feed by severing stems from the plant crown, first causing small yellow-brown spots along the edges of fairways, tees and collars of greens. Moderate infestations cause small irregular patches of dead turf and heavy attacks (up to 500 per ft^2) kills turf in large areas. Damage begins to become obvious in late May or early June and is occasionally attributed to other causes such as disease. Damage on greens and tees may appear patchy because larvae feed on *P. annua*, not bentgrass.

The earliest recognizable symptom of larval damage is **general yellowing of Poa annua patches**. Upon close examination with a 10X hand lens, evidence of larval burrowing in the stems, or the small larva itself may be seen.

Symptoms of damage are often first seen on the edges of greens, tees and fairways. Yellowing usually spreads as the season progresses. Damage is most obvious during May and June. Examination of the thatch and soil in heavily damaged areas reveals larvae, pupae and light-brown adults. The hand lens is useful in detecting the **characteristic U-shaped notches made by the larvae at the stem bases**.

Life Cycle and Habits.
The annual bluegrass weevil overwinters as an adult in protected areas, such as golf course roughs, and can be found among the leaves and debris under trees and shrubs, especially under white pines. As the spring thaw begins, adults become active, fly or crawl to nearby annual bluegrass and begin feeding. Adults remain hidden in thatch during the day, but climb on grass blades and often fly to lights at night.

During April and May, the female chews a small hole through the leaf sheath and deposits eggs in groups of 2 to 9. Upon hatching, the larvae feed by tunneling in the stem, causing **death of the central leaf**. The larvae then move to the crown and begin feeding at the base of the grass plant by chewing out characteristic U-shaped areas in the crown. Older larvae establish small burrows in the thatch from which they emerge to feed at the base of nearby stems.

Mature larvae burrow 1/4 inch into the soil beneath the thatch, form a small cell and transform into the creamy-white pupal stage. After about eight days, the adults emerge. Studies in New York have shown that there are two generations each year. Second generation larvae are present from mid-August into September. This second generation rarely damages the turf to the same extent as the spring generation. Adults migrate to overwintering sites during September and October.

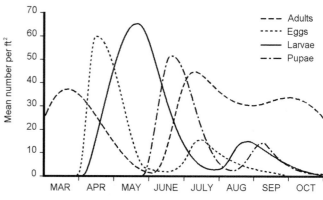

Seasonal occurrence of annual bluegrass weevil in New York.

Cutworms and Armyworms

Several species of thick bodied, non-hairy caterpillars in the cutworm-armyworm family may damage turfgrasses. The black cutworm is a primary pest of golf course greens, tees and fairways throughout the United States. This species rarely causes damage to lawns. The armyworm, fall and yellowstriped armyworms also occasionally damage golf course turf but are commonly associated with damage to home lawns. The bronzed, variegated, and glassy cutworms are principally pests of home lawns. The glassy cutworm is common in Canada.

The black cutworm, armyworm and fall armyworm are native to North America but they have been spread world wide by accidental introductions. The black cutworm and fall armyworm are actually tropical and semitropical species that fly from the Gulf States to cool-season turf areas each spring. The armyworm, bronzed, variegated, and glassy cutworms can survive northern winters.

All species of turfgrasses are hosts to these larvae. Black cutworms and fall armyworms most commonly damage the short cut turf found on golf course greens and tees.

Adults. The adults are dull brown and gray colored moths with wing spans of 1-3/8- to 1-3/4-inch (35 to 45 mm). At rest, the wings are folded flat over the abdomen.

Larvae. The larvae have generally hairless bodies except for a few scattered bristles. Besides the three pairs of true legs, these larvae have five pairs of fleshy prolegs on the underside of the abdomen. Most cutworms have characteristic markings on the head and body that aid in species identification. Full grown cutworm and armyworm larvae are 3/16-inch (6.0 mm) wide and 1-1/4- to 2-inches (32 to 50 mm) long. Most

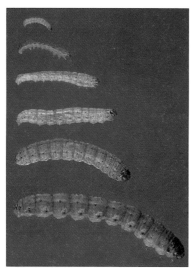

Black cutworm larvae, first through fifth instar.

cutworms and the armyworm coil into a spiral when disturbed.

Cutworms are so named because of their nocturnal feeding habit of cutting off plants close to the ground. On golf course greens and tees, black cutworms and fall armyworms

Black cutworms usually coil tightly when disturbed.

graze on the grass blades of short cut turf causing **circular or finger-shaped sunken areas, similar to ball marks**. Armyworms feed on grasses any time of the day and are known for their habit of moving and feeding, en masse, from one turfgrass area to another. They commonly eat everything green, leaving only a few stems. Bronzed cutworms occasionally damage cool-season turf under the cover of snow. Other species of cutworms are relatively uncommon and their damage is minor.

Detection and Monitoring. Regular monitoring of turf areas for evidence of cutworm or armyworm infestations and applying treatment only when larvae are present and/or damage seems eminent, are key to the curative approach to control. Monitoring includes looking for larvae, damage and/or evidence of birds (starlings) probing the turf for larvae.

Starlings commonly feed on cutworms, armyworms and sod webworms in all kinds of turf.

They leave characteristic probe marks in the turf.

To determine if larvae are actually present, an effective method is to use a flushing solution of liquid soap and water (two tablespoons of liquid Joy® dishwashing detergent in two gallons of water) spread over a one square yard area to flush larvae to the surface. In our experience, this solution has not damaged turf.

Life Cycle and Habits.
Black cutworms and fall armyworms overwinter as larvae or pupae only in the southern states and migrate to the northern states in early spring as moths. Armyworms overwinter as larvae or prepupae in the northern states. In the northern states, these insects may have two to four generations per year while in the southern states, three to seven generations may be found depending upon the length of the season. The bronzed cutworm overwinters as an egg or larva and has a single generation per year.

Adults of most cutworms and armyworms mate and feed at night on the flowers of trees, shrubs and weeds. Mated females of black and bronzed cutworms seek out crops or grasses and lay single or small clusters of eggs on leaf blades. Armyworms and fall armyworms usually lay several clusters of eggs on host plants or on structures (flag or light poles, overhanging trees and shrubs, etc.).

Large black cutworms excavate holes into the thatch and soil and venture forth at night to feed on plant material. On short cut greens and tees, the larvae simply eat the green leaves down to the ground, leaving pock areas similar to ball marks. Fall armyworms also follow this behavior. Older armyworms do not hide during the day but feed continuously.

Black Cutworm

The black cutworm is a cosmopolitan pest found across North America, Europe, Asia, and Africa. It does not survive in areas that regularly get below 15°F during the winter. Larvae feed on all species of turfgrasses and are serious pests of corn, vegetables, and other agricultural crops.

Adults are generally dark gray to black colored moths with some brown markings. The forewings span 1-3/8- to 1-3/4-inch (35 to 45 mm) and a black dagger-shaped mark appears at the outer edge.

Adult black cutworms feed on flowers at night, mate and lay eggs on turf.

The larvae have generally hairless bodies except for a few bristles scattered over the body. The general upper body color is gray-green to nearly black and slightly lighter gray undersides. Often, a pale mid-dorsal stripe is visible. Under a microscope, the body integument appears to have a **cobblestone surface**. Mature larvae are 1/4-inch (6 mm) wide and 1-3/16- to 1-3/4-inches (30 to 45 mm) long.

Damage and Diagnosis.
The black cutworm is a true cutworm and has semi-subterranean habits. They usually burrow into the thatch/soil or use existing cracks and crevices or aeration holes, from which they emerge at night to clip off grass blades and stems. This feeding damage often appears as **circular depressed spots that resemble ball marks** on golf greens. Damage may also appear as elongated or glove shaped pock marks.

Typical circular feeding spot of black cutworm that resembles a ball mark.

Application of the soap flush solution previously described for sod webworms is effective for diagnosing infestations and assessing the effectiveness of treatments against the larvae. Birds (especially starlings) visit infested areas frequently and leave **probe marks** in the turf. On golf greens, **tufts of turf** are left on the surface when the birds remove larvae.

Life Cycle and Habits.
This insect overwinters in southern states (probably as larvae and pupae). The adults fly and are blown northward with spring weather fronts, arriving in Iowa to Ohio about mid-April. These adults are believed to be the source of larval infestations in turf in northern and mid-western states.

Migrating females are mated when they arrive. Adults actively feed on nectar from flowering plants at night. Each female has the capacity to lay 1200 to 1600 eggs over 5 to 10 days. In turf, black cutworms attach eggs to the tips of grass blades. Eggs are usually laid singly. The eggs hatch in 3 to 6 days, and the first instar larvae feed on turf leaf surfaces. After several days and molting, the larvae work their way to the base of grass plants.

Cutworm larvae excavate burrows into thatch and soil. On golf courses, the larvae occupy aeration holes or make burrows in the holes started by spikes on golf shoes. The larvae venture forth at night from these burrows to feed on turf leaves and stems, leaving pock marks on greens and tees. Most of the larvae complete 6 to 7 molts in 20 to 40 days. The pupa is usually formed in the cutworm burrow and takes about two weeks to mature. Just before the adult emerges, the pupa wriggles its way to the thatch surface. This motion is often detected by birds who probe for the pupae.

Developmental times can be greatly lengthened during the cooler months, but generally take 40 to 65 days. There are two generations in New York, three in Ohio, Missouri and the Central Great Plains, four in Tennessee and five to six in Louisiana. Subtropical regions probably have continuous generations.

Black cutworm adults attach eggs to tips of grass blades.

Armyworm

The armyworm (=common armyworm) is found throughout North America, generally east of the Rocky Mountains, though occasionally in Utah, California and British Columbia. It feeds almost exclusively on grasses, but in the absence of grasses, larvae have been known to damage vegetables, ornamental plants, as well as cereal and forage crops.

The adults are light tan with darker shading. They have a distinct, small, light colored, diamond-shaped spot on the mid forewing. Adults are about 3/4-inch (20 mm) long and have a 1-3/16-inch (30 mm) wing span.

Armyworm larvae are brown to greenish with distinct yellow and brown stripes, two of which are **narrow and broken**. The head capsule is yellowish-brown with a broadly **H-shaped pattern**. The larval body generally tapers from front to back.

Armyworm adult on grass stem.

Damage and Diagnosis.
Armyworms feed on grasses any time of the day and are known for their **habit of moving and feeding**, **en masse**, from one turfgrass area to another. They commonly eat everything green, leaving only a few stems. This "moving front" of larvae may consume many square feet of turf which seems to disappear overnight. Large populations may also develop in grassy meadows and along rights-of-way for road and utilities. Larvae tend to **curl up when disturbed**.

Life Cycle and Habits.
The armyworm overwinters as a mature larva or pupa in the southern and central regions of the United States. In northern states, some larvae may overwinter, but adults also fly or are blown north with spring weather fronts. Larvae that overwinter pupate and emerge in late March and April (Ohio). These adults and the migrating females lay clusters of 100 to 300 eggs on host plants or on structures (flag or light poles, overhanging trees and shrubs). Females are capable of laying several thousand eggs. When the larvae first emerge, they stay together feeding on the same plant until everything is devoured. The larvae feed on grasses any time of the day and go through 6 to 9 instars over a period of 20 to 48 days (depending on temperature) before pupating in thatch and soil.

Mild winter temperatures combined with moderate, moist spring and early summer conditions favor armyworm outbreaks. In Ohio, the second (June and July) or third (August) generations are most likely to cause damage in turf. Based on the occurrence of adult flights, there appears to be two generations in New York, three in Ohio, and 4 to 5 in North Carolina and south from Tennessee.

Mature armyworm larva on grass stem. Note distinct stripes and H-shaped pattern on head.

Ohio lawn damaged by armyworms in July. Note neighbor's lawn that was treated by a lawn care company that had discovered the armyworms.

Fall Armyworm

The fall armyworm (and its near cousin, the yellowstriped armyworm) is thought to be semitropical in origin, probably from Mexico or Central America. It is a permanent resident of the Gulf States and adults migrate north and westward during the spring and summer. Since it is not able to withstand freezing temperatures, northern populations die each winter.

The larvae feed on a wide range of host plants but prefer grasses such as bluegrass, ryegrass, bentgrass and fine fescue. Damage to bermudagrass, St. Augustinegrass, and other southern grasses is common, especially during warm, dry seasons.

The adult males and females have distinct differences in the markings on their wings. Both are generally shades of gray with white markings. The females can be almost entirely gray without much marking. Both males and females have a characteristic drop-shaped light mark in the middle of the forewing that trails towards the hind tip margin. They range from 11/16- to 1-inch (17 to 25 mm) long with wingspans of 3/4- to 1-3/16-inch (20 to 30 mm).

Fall armyworm male.

Mature fall armyworm larva showing Y-mark on head.

Fall armyworm damage to bermudagrass fairway and rough in Florida.

Mature <u>larvae</u> are 1-1/2-inch (30 mm) long and <u>have distinct stripes</u>; a faint line in the middle and two black lines along the sides. Each abdominal segment has four distinct spots. Body color ranges from pinkish, to light tan-yellow, to green, to nearly black. The **head has a characteristic, yellow-white, inverted Y-shaped mark**.

Damage and Diagnosis.
Fall armyworm <u>larvae feed anytime of the day</u> but are most active in the early morning and evening. Entire leaves and stems are consumed giving the turf a ragged appearance. If **touched, larvae coil tightly**. When numerous, **larvae may move**, **en masse**, from damaged areas to fresh turf. Occasionally, severe infestations may cause permanent damage. Late summer migrations have been known to produce black cutworm-like damage to golf course greens in September.

Life Cycle and Habits.
This pest is a <u>constant threat to southern turf</u> and it <u>rarely damages transition or northern turf</u> because it can not overwinter. All stages are present throughout the year within 100 miles of the Gulf Coast. Adults lay <u>egg masses on preferred grasses</u> and small grain crops, but <u>may lay egg masses on trees, shrubs and flags on golf greens</u>.

Fall armyworm feeding around aerification hole on green.

The eggs hatch in 7 to 10 days and can hatch in 2 to 3 days in July and August. Young larvae spin silken threads that help disperse them to their host plants. The larvae feed together until they reach the fourth or fifth instar when they may become cannibalistic. Larger larvae feed during the day and if disturbed by a shadow or touching, they drop from the leaves and coil tightly. The larvae can take as little as 12 days to mature in July and up to 28 to 30 days in October. Sixth instar larvae pupate in the lower thatch and take 9 to 20 days to mature.

<u>Adults emerging in the southern states migrate north, south and west with weather fronts</u>. Generally, one generation is completed by the northern migrants and three or more by those moving south and west. Larvae from the northern migrants have been known to cause damage to golf greens similar to that from black cutworm in September. However, their preferred hosts are corn and soybeans.

Fall armyworm eggs attached to St. Augustinegrass leaf blade.

Fall and yellowstriped armyworm adult commonly attach their eggs to golf course signs and the hole marker flags.

Yellowstriped armyworms commonly occur with fall armyworms in southern turf.

photo: H. Tashiro

The lawn armyworm behaves much like fall and yellowstriped armyworms, but it is found in Hawaii.

Bronze Cutworm

The bronze cutworm generally occurs east of the Rocky Mountains. Larvae are most commonly found in bluegrass home lawns but can be found in golf course roughs and pastures.

The <u>adults</u> are purple-gray to brown with a wide dark-brown band across the center of the front wings. Adults are found in September and October.

The upper body of the large <u>larvae</u> is generally light to dark brown with a distinct bronzy sheen and broad yellow stripe down the middle. **Behind the head is a dark brown collar with three white bars**. The first three instar larvae are green.

 photo: J. Fenstermacher

Extensive damage to New England lawn in late March, caused by bronze cutworm feeding under snow during winter months.

Mature bronzed cutworm.

Damage and Diagnosis. The larvae feed deep into plant crowns and can occasionally consume considerable areas of turf. In the spring, robins are often seen pecking larvae from home lawns. During mild winters, especially with snow cover, larvae can consume most of the turf foliage and after the spring melt, the lawn looks like it has been severely damaged by snow mold or winter kill. <u>Larvae will accumulate under a piece of plywood or cardboard placed on the turf overnight</u>.

Life Cycle and Habits. Generally, <u>this species overwinters as eggs</u> laid by adults that appear in September and October. However, eggs may hatch during warm winter periods and the larvae can feed and damage turf under the cover of snow in February and March (Shetlar observation). Spring hatching larvae mature by early to mid-June, dig into the soil and pupate, and remain dormant for the summer. There is only one generation per year.

Second & third instar bronzed cutworm larvae (left) and mature larva (right).

Other Turf, Leaf-Feeding Caterpillars

Lawn Armyworm

photo: H. Tashiro

The <u>lawn armyworm occurs on Pacific Islands</u>, including Hawaii, but <u>is not known in North America</u>. The larvae feed on a wide variety of grasses but bermudagrass and zoysiagrass are preferred hosts. The larvae are green with brown to black stripes and black dashes next to a yellow stripe. As with other armyworms, **larvae are known to move, en masse**, devouring grass blades as they move. Bermudagrass home lawns are damaged most. Continuous generations occur and population outbreaks are often related to rainy seasons.

Lawn armyworm damage to large lawn area in Hawaii.

Striped Grassworm (= Grass Looper)

The striped grassworm is a <u>periodic pest from Texas to Florida</u>. Larvae feed on the blades of bermudagrass and St. Augustinegrass. Heavy infestations can completely strip all foliage, leaving only stolons. Light infestations leave the turf ragged.

The larvae have <u>striking brown stripes</u> from the head to the tip of the abdomen. They are typical "loopers" in that they <u>move as</u> <u>inchworms over surfaces</u>. Several generations occur each year but the late summer and early fall populations do most of the damage.

Mature grass looper larva.

Fiery Skipper

The fiery skipper is <u>actually a butterfly</u> that occurs over most of North and South America, is most abundant in the Gulf States, and has been found in Hawaii. Larvae prefer bermudagrass but St. Augustinegrass is also commonly fed upon. Other skippers also feed on grasses and their larvae can occasionally be found in turf from Canada to Mexico.

<u>Adults</u> are <u>robust yellow butterflies</u> with orange and brown markings on their wings. They are often seen setting on golf greens during the day.

The <u>larvae</u> are unique in form, having dark heads, <u>an obvious constriction at the neck</u> and plump, yellow to gray-green bodies without stripes. The <u>body is covered with very tiny hairs</u>. The <u>solitary larvae spin loose silk shelters in the turf canopy and emerge from it to feed on turf leaves, creating round spots, 1 to 2 inches in diameter</u>. These spots may join to form larger damaged areas. Turf with considerable crabgrass seems most prone to injury. Three to five generations may occur.

Adult fiery skippers feed on flowers.

Fiery skipper adults attach eggs to turf leaves.

Fiery skipper larva showing characteristic constriction behind head.

Chinch Bugs

Chinch bugs are "true bugs" and attack turfgrasses in North America. The hairy chinch bug is the primary pest of northeastern turfgrasses while the common chinch bug is more commonly found in the northern Plains States and transition turf zones. The southern chinch bug is a pest of warm-season grasses. These species are difficult to separate in the field. Locality and host plants are criteria for identification. A fourth species has been recently described that feeds primarily on buffalograss.

Hairy Chinch Bug

The hairy chinch bug's range extends throughout the northeastern states and all Canadian provinces in and east of Ontario. The common chinch bug shares part of the north-central range.

Hairy chinch bug's preferred hosts include fine fescues, perennial ryegrasses, Kentucky bluegrass, bentgrass and zoysiagrass. The common chinch bug feeds on these grasses as well as grain crops such as sorghum, corn and wheat.

Hairy chinch bug <u>adults</u> are approximately 1/8-inch (3.5 mm) long and 3/64-inch (1.8 mm) wide, males being slightly smaller than the females. The head, pronotum and abdomen are <u>gray-black in color</u> and covered with fine hairs. The <u>wings are white with a black spot located in the middle front edge</u>. The legs often have a dark burnt-orange tint. Individuals in a population, or in some cases, most of a local population may have <u>short wings</u> that extend only half way down the abdomen.

There are five <u>nymphal instars</u>, each of which <u>change considerably in color and markings</u>. The first instar has a **bright orange abdomen with a cream colored band**, brown head and thorax and is about 1/32-inch (0.9 mm) long. Second through fourth instars continue to have this general color pattern except that the orange color on the abdomen gradually changes to a purple-gray with two black spots. The fourth instar increases to more than 3/32-inch (2 mm). In the fifth instar the wing pads are easily visible and the general color is black. The abdomen is blue-black with some darker black spots and the total body length is about 1/8-inch (3 mm).

(left to right) **Hairy chinch bug stages: egg, 1st, 2nd, 3rd, 4th, 5th instar nymphs, winged and short-winged adults.**

Damage and Diagnosis. Chinch bugs generally <u>occur in scattered patches</u> rather than being evenly distributed over the turf. Sunny areas are most heavily infested with populations sometime reaching 200 to 300 per square foot. Plant injury occurs as a result of the insect sucking fluids from the plant and at the same time injecting salivary fluids into the plant. The presence of the salivary fluid disrupts the water-conducting system of the plant, causing it to wilt, turn yellow, then brown and die (i.e., <u>chinch bugs kill turf!</u>). ***Injury is particularly severe when heavy infestations occur in turf that is dormant from moisture stress.*** *Such* ***dry conditions are particularly conducive to chinch bug growth and population development.*** <u>Visual scanning of sidewalks and driveways</u> adjacent to infested turf on hot afternoons often reveals <u>adults running across the pavement</u>.

Several techniques work for detecting or monitoring chinch bug populations in turf. The simplest method is the "**hands and knees**" method. Use your thumbs and fingers to pull back the grass stems to expose the crowns, thatch and chinch bug adults and nymphs that hide at the base of the plants. <u>Early stage nymphs are</u>

Chinch bugs prefer sunny sites. Damage first appears as irregular patches of drought stricken turf. Adults are often seen running across sidewalks on hot days.

If left uncontrolled, chinch bugs can kill extensive patches of turf.

First instar hairy chinch bug nymphs insert their sucking mouthparts as soon as they emerge. Note white band.

Using the flotation technique to detect chinch bugs.

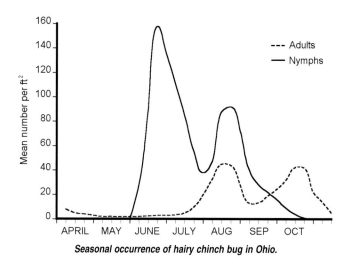

Seasonal occurrence of hairy chinch bug in Ohio.

very small and easily overlooked. During very hot and dry conditions, the chinch bugs may be located deep within the thatch.

A reliable detection method is the "**flotation technique.**" Cut the lid and bottom from a two-pound coffee can, and trim off the bottom rim to produce a sharp edge. Twist the sharp edge of the can through the turf into the underlying soil. Fill the can with water and count the chinch bugs that float to the surface in 10 minutes. Refill if the water soaks into the ground before the 10 minute period ends.

Life Cycle and Habits. Hairy chinch bug adults

overwinter in thatch, clumps of grass, along the edges of sidewalks, and next to buildings. Aggregates of 300 or more per square foot are often found. Adults become active when the daytime temperatures reach 70°F. The females feed for a short period, mate and begin laying eggs by inserting them into the folds of grass blades or into the thatch. This usually occurs from mid-April into June, from New York to Illinois. A single female may lay up to 200 eggs. Eggs

take about 20 to 30 days to hatch at temperatures below 70°F but can hatch in as little as a week when above 80°F.

Upon emerging from eggs, the nymphs immediately begin to feed by inserting their mouthparts in grass stems, usually while under a leaf sheath. Usually, the first generation matures by mid-July, when on hot summer days, considerable numbers of adults and larger nymphs can be seen **walking about on sidewalks or crawling up the sides of light colored buildings**. Damage may be visible from late June through August when the mature summer generation nymphs and adults are feeding. Summer females lay eggs and second generation nymphs mature from the end of August into September. Late nymphs die before winter but adults seek out protected areas to overwinter, but they also overwinter in turf.

Southern Chinch Bug

The southern chinch bug is found from southern North Carolina to the Florida Keys, west to central Oklahoma, California, and Hawaii. This species occasionally attacks centipedegrass, zoysiagrass, bahiagrass and bermudagrass, but is a major pest of St. Augustinegrass wherever it is grown.

Five to six nymphal instars occur, each of which vary in color and markings. First instars are bright orange with a cream colored band across the abdomen. Second through fourth instars continue to have this same general color pattern, except that the orange color of the

abdomen changes to a dusky-gray with small black spots. Wing pads of fifth instars expand and are easily visible. The abdomen becomes blue-black with some darker black spots.

Adults are gray-black to dark chestnut-brown and covered with fine yellow to white hairs. The wings are white with a black spot in

Southern chinch bug adult and large nymph feeding on St. Augustinegrass.

70

the middle front edge. The legs are chestnut-brown as is the hairy chinch bug. Individuals in a population may have short, nonfunctional wings which reach only halfway to the abdomen.

Damage and Diagnosis.
Infestations usually cause irregular, expanding patches of St. Augustinegrass or bermudagrass to turn yellow and then brown. As populations increase, extensive areas in a lawn may be killed, and the infestation may **move en masse**, across sidewalks and driveways to another area. Adults sometimes congregate on the tips of other grasses and weeds. Populations of over 1,000 chinch bugs per ft^2 have been recorded in heavily damaged St. Augustinegrass.

Adults and later stage nymphs are **commonly visible on sidewalks and driveways** adjacent to infested turf. **The water flotation method previously described is also useful for detecting infestations and measuring the effectiveness of control treatments**.

Life Cycle and Habits.
In its southern range, adults and a few nymphs in various stages overwinter in St. Augustinegrass and bermudagrass, especially in thatch. In the northernmost part of the range, only adults overwinter. These insects may become active any time the temperatures rise above 65°F but reproduction generally does not occur until April or May.

Females deposit their eggs by forcing them between the leaf sheath and stem, and occasionally in thatch. Each female may lay 45 to 100 eggs over several weeks. Eggs hatch in 8 to 9 days at 83°F and 24 to 25 days at 70°F. Newly emerged nymphs immediately

Southern chinch bugs cause wilting and yellowing of St. Augustinegrass. When chinch bugs are not controlled, the turf will eventually die.

begin feeding under the leaf sheaths. Nymphs emerging elsewhere crawl into available spaces under leaf sheaths. Nymphs may feed in aggregates. Five and occasionally six instars occur over 40 to 50 days during warm weather. In cool weather, they may take 2 to 3 months to mature. In most Gulf States, 3 to 5 overlapping generations occur each season. However, 7 to 10 generations occur in Southern Florida and the Caribbean Islands.

After overwintering, the first two generations are fairly well defined. The first major adult peak usually occurs in June, and the second in August. Peaks from subsequent generations may occur in October and December. Most damage occurs during the summer dry season.

Other Thatch Inhabitants

Research has shown that turfgrass thatch is inhabited by a multitude of insects, mites, spiders and other arthropods. Studies in Ohio have recorded populations of over 5,000 insects and mites per square foot. The following are a few additional pests and non-pests that commonly inhabit turfgrass thatch.

Bigeyed Bugs.
Bigeyed bugs are important **predators** of chinch bugs and other insects found in the turfgrass environment. Some 20 species occur in the United States. They frequently occur in turf infested with chinch bugs and actively feed on all stages. Occasionally, this predator is so effective in removing the chinch bugs that only bigeyed bugs remain. In such cases, the damage seen is often mistakenly associated with the bigeyed bug. While bigeyed bugs have occasionally been observed feeding on turfgrass, their primary source of food is insects.

Chinch bugs and bigeyed bugs are somewhat similar in appearance but can be distinguished by the fact that the body of the chinch bug is narrow, the head small, pointed, triangular shaped, with small eyes; while the body of the bigeyed bug is wider, the head is larger, blunt, with

Bigeyed bugs feed on chinch bug nymphs and other small insects.

two large, prominent eyes. Bigeyed bugs run quickly over the turf surface and are much more active insects than the slower moving chinch bugs.

Mites.
Mites are not insects, in fact they are close relatives of spiders (they have eight legs). Most species that inhabit thatch are beneficial, feeding on organic matter or preying on insects and other arthropods. Pest species hide in the thatch and crawl up the plant to feed on the leaves. If present, these mites are often crushed during employment of the "hands and knees" method of detection, leaving the observer with green-orange stains on his hands and knees. Two of the species that cause damage are shown here - winter grain mite and clover mite. The biology and damage caused by these mites are included in Chapter 6, Leaf and Stem-Inhabiting Pests.

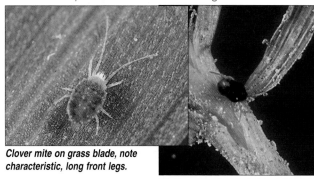

Clover mite on grass blade, note characteristic, long front legs.

Winter grain mite has red-orange legs and a dorsal anus.

Ground beetles are common beneficial predators found in turf. However, many people do not "appreciate" these beneficials and would prefer not to encounter them.

Ground beetle larvae are also beneficial predators though some are known to feed on seeds and plant roots.

Millipedes generally feed on decaying organic matter and are considered beneficial. When large numbers occur, millipedes often enter buildings where they are considered a nuisance pest.

Springtails are usually round or elongate. All have the ability to jump when disturbed, and are often mistaken for fleas. They are normally numerous in turf thatch where they feed on organic matter, molds and small organisms.

Bibionid or March fly larvae (curved individuals in picture) are commonly found feeding on decaying thatch and in small patches of turf killed by disease. The larvae are not considered pests. Prepupae (far right) and pupae (far left) may also be found in the thatch.

March fly adults are commonly called "lovebugs" because they fly joined together.

Other Turf-Infesting Billbugs

Sphenophorus coesifrons has no common name, but it is commonly found in warm-season turf, especially bermudagrass and St. Augustinegrass. It is a large black species, similar in size to southern forms of the hunting billbug. It has a characteristic, smooth raised mark in the middle of the pronotum. It looks like a very large bluegrass billbug. Nothing is known about the life cycle of this billbug, though adults are commonly found in the turf in the fall, winter and spring.

Sphenophorus minimus is often called the lesser billbug though it has no official name. It is very common in the eastern half of North America where cool-season turf is cultured. It is often found in populations of bluegrass billbugs and occasionally is more common than bluegrass billbug. It is generally smaller than the bluegrass billbug and the pronotum is covered with a mixture of large and small pits. The wing covers also appear rough with large and small pits (the bluegrass billbug wing covers have smooth rows of pits). The seasonal life cycle appears to be the same as the bluegrass billbug.

Sphenophorus apicalis, often called the apical billbug, has two distinct pits on the pronotum, just behind the head. It is smaller than the hunting billbug and can be common in southern and transition zone turf. Nothing is known of its hosts or life cycle.

Notes

Notes

Chapter 6

Leaf & Stem-Inhabiting Pests

This Chapter includes those arthropods (insects and mites) that feed on the upper leaves and stems of turfgrass plants. Many of these pests often hide in thatch, others remain exposed on leaf surfaces, and the rest hide in the spaces beneath leaf sheaths and nodes. Most of these pests have piercing-sucking or rasping (mites) mouthparts that pierce the plant and withdraw plant liquids as food. While this alone causes plant stress (yellowing and loss of turgidity), the primary cause of plant death from such pests is that the feeding process includes injection of salivary fluids into the host plant. By various means, including plugging vessels that translocate water and nutrients, these substances can cause the plant to die. The greenbug aphid, mealybugs and twolined spittlebug are included in this group.

Mites

Mouthparts of mites are generally of the piercing-sucking type. Mites use these mouthparts to probe surface plant cells and then suck up the fluids as food. This feeding activity, plus the drying effect of sun and wind blowing across the plants, causes the turf to appear as though suffering from moisture stress. There is some evidence that toxic substances may also be injected into the plant.

Bermudagrass Mite

The bermudagrass mite occurs in Australia, New Zealand, and probably wherever bermudagrass is grown. In the United States, it is found in the southern states where bermudagrass, its only host, is grown.

Characteristic witches'-brooming of bermudagrass stems caused by bermudagrass mites.

This mite has stages typical of the eriophyid mite group. They are extremely small with adults 0.006-inch (0.2 mm) long, pear-shaped with wormlike, soft bodies, and only two pairs of short, forward-projecting legs. A 20X hand lens

Bermudagrass mites exposed by pulling back leaf sheath.

or microscope is needed to see them. Interestingly, only females are known.

Damage and Diagnosis. Damage is first noticed when bermudagrass does not have vigorous growth in spring and is often yellowed. The turf appears stunted, and close inspection reveals that the stem length between nodes is greatly reduced. *Leaves and buds become bushy, forming a rosette or*

turf which is called "witches'-brooming." Heavy infestations produce an open, "tufted" appearance with irregular sections eventually turning brown and dying. Damage is most severe during hot dry weather.

Life Cycle and Habits. Because of their small size, these mites are very difficult to study, and little is understood about their life cycles and habits. Most eriophyid mites of this type lay less than a dozen eggs during their adult life and these usually hatch in 2 to 3 days. At 75°F, it is estimated that adulthood is reached in 7 to 10 days, and eggs are laid for 2 to 5 days. A cycle can be completed in 10 to 14 days. This short time period allows for a rapid buildup of a population during summer temperatures.

The bermudagrass mite is <u>tolerant of high temperatures</u>, having only moderate mortality at 120°F. Cold temperatures tend to stop development, though surviving during the winter only occurs where bermudagrass remains green at the soil surface. This mite can be <u>spread by the wind or carried on the bodies of other insects</u>. The most common method of spread is by transportation with infested sod. The mites can not survive on bermudagrass seed.

Banks Grass Mite

The <u>Banks grass mite is a spider mite</u> originally described from the Pacific Northwest, but is now known to occur from Washington to Florida and south. It occurs in Hawaii, Puerto Rico, Central America, Mexico and Africa. This mite commonly attacks Kentucky bluegrass in Washington, Oregon and Colorado, but infests bermudagrass and St. Augustinegrass in the southern states.

The first stage immatures (larvae) have only three pairs of legs. The next two stages (nymphs) as well as adults have four pairs of legs and are bright green as they feed on turf.

Adult females are broadly oval and about 0.016-inch (0.4 mm) long, while males have strongly tapered abdomens and are only about 0.013-inch (0.3 mm) long. During the spring to fall feeding periods, the adults are bright green with light orange legs, but during winter, the green fades and the mites are a bright orange-salmon color.

Damage and Diagnosis. Lightly infested plants have small yellow speckles along the leaves. As damage progresses, the leaves become more straw colored and they eventually wither and die in hot, dry weather. ***Damage to warm-season turf may be mistaken for summer dormancy***. This mite overwinters in all stages and during warm winters large numbers of mites may cause damage by the following spring. Spring damage on cool-season turf may appear as a <u>ring of desiccated turf surrounding coniferous trees or as patches of winter desiccated turf around buildings</u>. ***Damage to cool-season turf may be mistaken for snow mold or clover mite injury.*** This mite produces considerable <u>webbing at the bases of turf tillers</u>. These <u>webs are easily visible in the morning dew</u>. In St. Augustinegrass, the mite egg shells and webbing are sometimes mistaken for molds or dust. Concentrations of the mites on the tips of southern grasses cause general yellowing and <u>dieback similar to heat or drought scorch</u>.

Banks grass mites.

Banks grass mite damage to cool-season turf may appear similar to clover mite, snow mold or winter desiccation.

photo: W. Cranshaw

Life Cycle and Habits. Banks grass mite is most active at summer temperatures, but it can also damage turf under the cover of snow. Mated females overwinter at the base of grass plants and in the soil, but a few males and nymphs may also be present. <u>Overwintered females</u> lose their green color and become <u>orange-salmon</u>. In spring, surviving mites begin to feed on emerging grass and become green. Females lay 50 to 70 eggs in the dense webbing they produce on plants. The eggs take from 4 to 25 days to hatch depending upon the temperature. At temperatures above 70°F, the larval stage takes two days, and later stages 1 to 2 days each to mature. During hot weather, development from egg laying to adult may take as few as nine days. During hot dry summer weather, immature mites and sometimes adults migrate to the center of dormant grass clumps and rest until the grass returns to active growth following rains. During cool spring and fall temperatures, complete development may take 25 to 37 days. This means that 6 to 9 overlapping generations may occur in a season.

As soon as adult females appear, males begin copulation. If no males are present, the females can lay unfertilized eggs that develop only into males. These males can mate with the female (their mother) and she can then produce female offspring. Only mated females can produce female mites.

Webbing in bermudagrass from Banks grass mites.

Banks grass mite webbing visible in the morning sun. Inspect the turf carefully to confirm that the webbing is not the result of spring or fall spider hatchlings.

Clover mites in thatch.

Clover mite showing long front legs and red eyes.

Typical turf damage from clover mite feeding next to house foundation.

Clover Mite

The clover mite is a cosmopolitan species found in North and South America, Europe, Asia, Africa, and Australia. This pest attacks a wide variety of plants, including Kentucky bluegrass, ryegrass and clover.

Only females are known. Adults are reddish- to chestnut-brown, 0.016-inch (0.4 mm) long, and **have the front legs about twice the length of the other legs**.

Damage and Diagnosis. Like other mites, the clover mite probes the surface of grass blades giving a silvery appearance to the upper surface. Populations of four to five thousand mites per square foot are common. Sun and wind further desiccates host plants. **Damage is similar to and may be misdiagnosed as winter desiccation caused by wind**. Damage occurs on home lawns, usually next to buildings.

The major problem with these mites is their nuisance activities. During population flushes in the spring and fall, they migrate into homes. The mites do not transmit any diseases, bite people or feed on house furnishings or food. However, when crushed, the mites leave orange-red stains on walls, furniture and clothing that are difficult to remove.

On golf courses, this mite occasionally invades shelters and stains the light colored clothing of players unfortunate enough to sit on them. They also may cluster on sprinkler system switch boxes where they coat the surface with cast skins and egg shells.

The most obvious identification characteristic of this mite is its **long front legs** which extend forward from the body when disturbed. These legs are visible with a 10X hand lens.

Life Cycle and Habits. In cool-season turf, this mite overwinters as spherical, bright red eggs laid on the walls of buildings, tree trunks, etc., and/or as adults in houses and other protected areas. The eggs hatch and overwintered adults lay eggs in early March and April. Eggs laid in late May and June remain dormant during hot weather and hatch in September to October. Adults feed actively, often on turf next to buildings. Two to three overlapping generations occur each fall and spring. In the warm-season areas, feeding adults, nymphs and eggs can be found throughout the winter.

Winter Grain Mite

The winter grain mite is a pest of grains west of the Mississippi but is widely distributed throughout North America. Grasses, including Kentucky bluegrass, fine fescues and perennial ryegrass, are also hosts but damage to legumes, vegetables and other plants has been reported.

No reference to this mite being a pest of turfgrass was found until 1968 when it was reported causing damage to red fescue and Kentucky bluegrass in New Jersey. Producers of turfgrass seed in Oregon considered the winter grain mite a pest since, at least, 1968.

H.D. Niemczyk observed significant damage on bentgrass golf course fairways in Pennsylvania (1977), on fairways and greens in

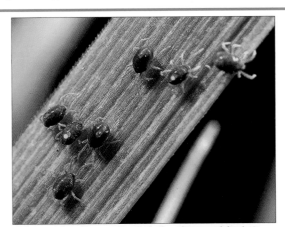

Winter grain mites crawl onto leaf surfaces and feed at night. Note dorsal anus that appears as a red-orange spot (except two with liquid droplets that appear yellow).

Winter grain mite eggs (salmon-colored, oval objects) attached to turf stem.

Lawn damaged by winter grain mites, April view.

Winter grain mite damage is often confused with winter desiccation.

Cincinnati, Ohio (1980), and a home lawn of Kentucky bluegrass and fine fescue in North Canton, Ohio (1979).

The adults are relatively large for mites, up to 3/64-inch (1 mm) long. They are the only turf-inhabiting mites with **olive-black bodies**, bright **red-orange legs** and mouthparts, a **pair of white eye spots**, and a **dorsal anus**. Only females are found.

Damage and Diagnosis.
During warm sunny winter days, this mite can be found on the crowns of grass plants, in thatch and at the soil surface. On overcast days and at dusk, the mites often "appear" in great numbers on the grass leaves. The **dorsal anus, surrounded by a red-orange spot** distinguishes this mite from all others one might find in cool-season grasses. While the mite is visible to the naked eye, a 10X hand lens is needed to see the dorsal anus. When disturbed, a droplet of liquid appears at the anal opening (probably an alarm pheromone).

Mite populations can increase to a peak of several thousand per square foot by late February or early March. The mites are most active at night and on dark cloudy days when, in seeming unison, thousands crawl from the thatch to grass blades and begin feeding. In feeding, mites probe cells on the leaf surface and consume their contents. Exposure of the surface cells to sun and wind leads to complete desiccation of the leaf. Except for the presence of mites, **symptoms of injury are the same as those associated with winter desiccation**; a condition caused by excessive transpiration accentuated by wind movement across exposed turfgrass. Symptoms are also similar to those from early spring frost injury. This similarity of symptoms has probably led to some misdiagnoses of winter grain mite damage.

Life Cycle and Habits.
The most distinctive feature about the winter grain mite's life cycle is the oversummering eggs and winter mite activity. In the northern United States, the mites appear to hatch in mid- to late-October when soil surface temperatures are approaching 50°F. The mites feed by probing the surface of grass blades and sucking up the cell contents. The mites tend to hide during daylight and can be found clustered in grass crowns, in thatch and at the soil surface during warm bright winter days. **Snow cover does not inhibit feeding and may actually afford protection**.

Females can live up to five weeks during which time they may lay 30 to 65 red-orange eggs glued to the base of stems or particles of thatch. Eggs soon shrivel and become darker with a white outer waxlike coating. Eggs laid from November through March usually hatch that winter. Eggs laid from March or later usually oversummer and hatch the following fall. It appears that two overlapping generations may occur during the winter with peak populations being found in late December and late February.

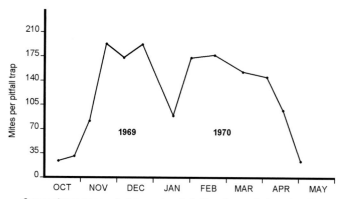

Seasonal occurrence of winter grain mite in New Jersey. *(redrawn from Streu & Gingrich, 1972)*

Bermudagrass Scale

The bermudagrass scale is found worldwide in tropical and subtropical regions. In the United States, it attacks bermudagrass from California to Florida and is known in Hawaii.

This scale is most frequently reported on bermudagrass, though it has been found on centipedegrass, bahiagrass, St. Augustinegrass and tall fescue.

Settled crawlers produce an oval, waxy test (shell) which is first straw yellow and later covered with white, waxy secretions. Adult female body shells are oval, white in color and 3/64- to 1/16-inch (1.0

Bermudagrass scales are located under leaf sheaths at nodes.

to 1.75 mm) long. Male scale tests are about one-half the size of females. Mature males are able to emerge from under their tests and are small gnat-like insects with one pair of wings. Their yellowish-pink bodies are about 0.02-inch (0.5 mm) long and have 2 to 3 long, white, waxy threads arising from the tip of the abdomen.

Damage and Diagnosis.

Bermudagrass first appears to **grow slowly, turn yellow in color, often resembling drought stress**. Heavy infestations may dramatically thin and kill patches of bermudagrass. This type of damage is more evident during periods of hot, dry weather. Where bermudagrass enters a winter dormancy, this scale can cause delay in spring green up.

When yellow turf is encountered, the scale should be sampled by digging out several affected stolons with attached above ground stems. Inspect nodes and bases of stems for oval, white scales. If the scale is confirmed, a sampling program should be followed to determine the extent of the infestation.

As settled scales grow, they begin to cover the body with loose waxy filaments which eventually give way to the formation of a solid, waxy shell-like test. At maturity, the scales often extend slightly from under the old leaf sheaths that originally hid the body. Populations can be so large at nodes and crowns that the scales seem to be stacked on top of each other.

Life Cycle and Habits.

Little is known about the actual time periods needed for development of this scale. Apparently

Bermudagrass scale damage on lawns and golf courses is often misdiagnosed as dry spots or disease.

the bermudagrass scale has two overlapping generations per year in southern states. In Georgia and Florida, eggs are laid and crawlers are most active during the spring rainy season. Settled crawlers and adults can be found during much of the season, though little growth occurs during winter dormancy or during summer drought periods. It is suspected that this scale has continuous generations in warmer climates where the bermudagrass does not become dormant.

Eggs are laid and retained inside the female scale shell. As eggs hatch, the tiny crawlers move along stolons and lower grass stems. They settle under old leaf sheaths at the bases of crowns, but may be found anywhere on stolons and lower stems. Often, large numbers of settled crawlers and young adults can be found on stems and stolons in the soil or thatch layer. These **scales are rarely exposed on upper plant parts**.

Rhodesgrass Mealybug (=Rhodesgrass Scale)

Rhodesgrass mealybug occurs from South Carolina to southern California and is found worldwide in tropical and subtropical regions of Africa, Australia, Central America, India, Japan, Pacific Islands, and South China. This species attacks over 70 species of grasses, including rhodesgrass, St. Augustinegrass, and bermudagrass where ever it is grown.

Only asexually reproducing females are known. The adult body is also saclike, broadly oval, dark purplish-brown and 1/16- to 1/8-inch (1.5 to 3.0 mm) long. The fluffy waxy covering turns yellow with age. The dark female body is exposed through openings at both ends of the waxy cover. A very long, 1/8- to 3/8-inch (3 to 10 mm), anal filament excretes a sweet liquid (honeydew). Though this pest is actually a mealybug, it is immobile, like a scale insect, once settled.

Damage and Diagnosis.

This mealybug does not commonly kill turf unless stressful conditions exist. High cut bermudagrass is more prone to severe damage during drought. **This pest produces considerable honeydew, therefore ants or bees may frequent heavily infested turf.**

Life Cycle and Habits.

The life cycle of the rhodesgrass mealybug continues throughout the year. Reproduction is considerably reduced during winter, but activity increases when bermudagrass begins rapid growth. Peak populations are reached in July, but are reduced during moisture stress in July and August. In September and October, populations again increase and peak in early November.

Mature rhodesgrass mealybugs showing characteristic anal wax tubes.

In spring, females lay an average of 150 eggs over 50 days. The eggs are contained inside the bodies of the female and her waxy cover. First instar nymphs (=crawlers) are flat, oval, cream colored with a median stripe tinged with purple, and have short legs and two waxy tail filaments. The crawlers settle on grass crown nodes, insert their piercing-sucking mouthparts, and begin secreting a waxy coat. After the first molt, nymphs take on a saclike form without legs. Only the threadlike mouthparts and anal excretory filament emerge from the body. A generation averages two months during summer months, but in winter may take 3.5 to 4 months to complete.

Apparently, this mealybug is moved by transportation on sod or grass clippings. Crawlers also crawl onto the legs of animals and thus may be moved about. High temperatures, especially near 100°F, reduce mealybug development and may actually kill individuals. Exposure of the mealybug to 28°F for 24 hours is fatal. Thus, winter cold limits northern movement and survival of this pest.

Overwintered greenbug eggs on dead truf (February, Ohio).

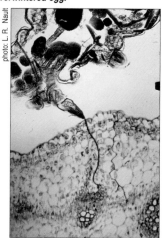

Greenbug nymph hatching from overwintered egg.

Greenbug small nymphs, wingless and winged adults.

Greenbug

The greenbug (**an aphid**) is reported to have damaged turfgrass from Kansas to New York, south into Kentucky and Maryland, and has been found in California. It is also a worldwide pest of cereal grains in Europe, Africa, and North America.

At least six biotypes have been identified and wheat, sorghum, oats, and over 60 members of the grass family are hosts. Host turfgrasses include Kentucky bluegrass, Canada bluegrass, annual bluegrass, fescues and perennial ryegrass. This aphid is known to reproduce on Kentucky bluegrass, chewings fescue and tall fescue.

Young aphids (nymphs) look like adults except smaller. Their pear-shaped body is light green and usually has a darker green stripe down the back. The tips of the legs, antennae, and cornicles (pipe-like structures on the upper side of the abdomen) are black. Nymphs destined to become winged forms have obvious wing pads in the last instar.

Adults are about 5/64 inch (2 mm) long and have the same green color and black markings as the nymphs. Winged forms usually appear when overpopulation occurs, often after considerable turf damage has occurred. Winged adults are usually darker green and have wing veins marked with black.

Damage and Diagnosis. Greenbug damage is

commonly found as circular yellowed areas under trees, but also occurs in open areas as well. There is only speculation about why the damage occurs under trees, but what is known is that **this aphid does not** (can not) **feed on trees**. Close (hands and knees) examination of turf in and around areas showing typical symptoms is necessary, since the aphids are not readily seen from a distance.

Young and mature greenbugs suck plant fluids from their grass host and simultaneously inject salivary fluids into the plant. This fluid causes the tissue around the point of injection to turn yellow, then orange. **Individual grass blades may have 50 aphids on them**. In heavy infestations, the **turf turns to a distinct burnt-orange color** (especially visible after rain or irrigation), and soon, **the complete plant dies**. Greenbug discoloration is often confused with the similar discoloration caused by turf rust disease. Greenbug infestations and damage can occur from June to November.

Life Cycle and Habits. The life cycle of the

greenbug in southern states (south of 35° latitude) is relatively simple. Without mating, wingless **females bear live female nymphs** that begin bearing more female nymphs in about seven days. A single female produces 30 or more young. Generations may be continuous.

Northern infestations originate from two sources: females that are blown north and east by spring weather fronts; and, females that hatch from **overwintered eggs** laid by females the previous fall. The wind blown females begin bearing live nymphs and the overwintered eggs hatch, also producing female nymphs. The females from both sources infest turf (and other grass hosts) and live for 20 days or so, producing 50 or more young each. Five to fourteen generations have been recorded in a season. Winged forms are produced when over crowding occurs, which facilitates establishment of infestations at other locations.

Photo of section through aphid (above), **showing stylets (mouthparts) extending to leaf phloem (below).**

Smaller winged males are produced in the fall. These males mate with females that lay overwintering eggs. The black eggs are glued to grass blades, fallen tree leaves, and other debris.

Characteristic orange-yellow-colored, greenbug infested Kentucky bluegrass turf under tree and in various spots of lawn. Note that fine fescue was not infested.

Twolined spittlebug mass around turf stem in thatch.   Twolined spittlebug nymph coaxed from mass.   Adult twolined spittlebug.

Twolined Spittlebug

The twolined spittlebug is a native of North America, most common in the Gulf States into Central America and is found as far north as Maryland to Kansas. This pest attacks southern turfgrasses, such as bermudagrass, St. Augustinegrass, bahiagrass and centipedegrass. Adults feed on herbaceous perennials.

First instar nymphs are pale yellow with a small orange spot on each side of the abdomen. Nymphs molt four times, during which time, the orange spots enlarge to cover the entire abdomen. The final, fifth instar nymph has well-developed wing pads with two transverse orange bands, and is about 5/16-inch (8 mm) long. The nymphs are usually covered by a "**spittlemass**."

The boat-shaped adults are about 3/8-inch (10 mm) long and have a dark brown to black color with **two distinct reddish-orange bands** on the wings. Adults have orange-red bodies, covered by the wings, do not produce spittlemasses, but readily jump and fly short distances when disturbed.

Damage and Diagnosis. Large numbers of
spittlemasses are unsightly and cause concerns by staining shoes and wetting bare feet. Nymphs cause patches of turf to yellow. Adults use their piercing-sucking mouthparts to probe leaves that turn brown and die. This gives the appearance of sparse, blighted turf. The adults may also cause damage to flowers or some ornamentals, especially Burford holly.

Life Cycle and Habits. Eggs overwinter in turf and
are usually deposited at the base of grass plants. Some eggs are inserted between a lower leaf sheath and stem, but more commonly into surrounding thatch. Eggs may be laid singly or in small groups. Though eggs are the normal overwintering stage, occasional adults may be found during the winter months in Florida.

Eggs hatch in early spring when the turf is recovering from dormancy. Newly emerged nymphs must find a suitable feeding site within an hour or two or they die. After probing several places, nymphs insert their piercing-sucking mouthparts into turfgrass tissues and produce a "spittlemass." This is actually an excretion from the anus. As feeding and development continues, nymphs may move about and sometimes several nymphs occupy a single spittlemass. If too much moisture is present in the turf, nymphs may move to the tips of grass blades to form spittlemasses.

Depending upon temperatures, nymphs take 34 to 60 days to mature. First peak adult emergence occurs in late may and early June in Florida to July in South Carolina. Depending on temperatures and moisture, adults may be continuously present until October, or a second peak of adults may occur from late August through September. Newly emerged adult females release a sex pheromone which attracts males. Mated females lay eggs in about a week and can produce about 50 eggs over a two week period. Eggs take 14 to 23 days to hatch in summer months, but eggs laid in the fall do not hatch until the next spring.

Frit Fly

The frit fly is a native of Europe where it commonly occurs in grain crops. The common name is derived from damage it causes to grasses and grain crops. The larvae (maggots) infest stems of wheat, oats and rye, eventually feeding on immature kernels of grain in the developing heads. Such heads produce empty kernels of grain called "frits." This insect is widely distributed throughout North America and is known to cause damage to bluegrasses, bentgrass, reed canarygrass and other grasses. Seed stems of bentgrass are often damaged.

The maggot-like larvae are 3/32-inch (2.0 to 2.5 mm) long and have no legs or head capsule. The anterior end is pointed and has a pair of tiny black hooks used for rasping food.

photo: M. Tolley

Frit fly maggot is exposed by pulling back leaf sheath. Note two black mouth hooks (on the right side of body).

Frt fly adult.

photo: M. Tolley

Adults are rather nondescript black flies with yellowish or white markings. They are about the size of the small flies that gather around decaying fruit, about 3/32-inch (2 to 2.5 mm) long.

Damage and Diagnosis.
The frit fly is probably responsible for more damage to golf course greens and collars than is realized. The adult fly is an annoyance to golfers on putting greens because they are **attracted to white objects, including golf balls and golf carts.** They are often seen hovering over golf greens and can be collected throughout the summer from mowed and irrigated bluegrass. The adults seem attracted to new shoots, feeding actively on exudates from them.

Turfgrass damaged by frit fly maggots has a general yellow (chlorotic) appearance at first and is **easily misdiagnosed as being caused by disease or other factors**. Close examination reveals the **central leaf of one or more shoots from the crown is affected, while surrounding shoots and leaves may be green**. As feeding progresses, the shoot dies. Careful examination at the base of infested shoots will reveal one or more white maggots, which when removed crawl actively. The two characteristic mouth hooks of the larva are visible with a 10X hand lens.

The apparent attraction of adult frit flies to white surfaces is useful in detecting their presence. If present and actively flying, adult flies will immediately land on a white handkerchief or golf ball rolled or dropped on the turfgrass.

Life Cycle and Habits.
Based on Ohio studies by Dr. Mike Tolley, larvae overwinter in a small tunnel eaten out of the grass stem. In the spring, when the grass resumes growth, the maggots feed by rasping inside the stem. As the maggots tunnel

Adult frit flies are attracted to white objects, often annoying and distracting golfers.

The first signs of frit fly larval activity is a central stem turning yellow.

downward, they may pass nodes, killing the stem from that point outward. The maggot matures in several weeks and forms a pupa inside the stem or in the duff surrounding the grass plants. They live for about one week. The flies can often be seen in considerable numbers resting on the tips of grass blades in the morning or evening sun and are known to alight on lighter colored surfaces such as golf balls or equipment placed on the turf. The new adults insert eggs in the space between a leaf and stem. Eggs take about a week to hatch in warmer weather.

In Ohio, peak adult populations occur in mid-May, late June, late July to early August, and mid-September. There are three to four generations in the northern states and four to five farther south.

Notes

Hunting billbug adult on turf surface. A target of INITIAL CONTACT.

Black cutworm emerging to feed at night. A target for FUTURE CONTACT & INGESTION

Greenbugs on leaf. A target for INITIAL CONTACT & INGESTION OF SYSTEMIC MATERIALS.

Chapter 7

Principles of Controlling Crown, Thatch, Stem & Leaf-Inhabiting Pests

The Target Principle

The objective of efforts to control insect and mite pests that inhabit crowns, thatch, stems and leaves of turfgrass plants (the **Target Zone**) is to eliminate the pests' ability to feed in these habitation zones and thereby prevent plant damage. Having the mind set to focus on reaching the pest in these zones is to apply the **Target Principle**.

Basis of Control with Insecticides

Control of crown, thatch, stem and leaf-inhabiting pests is less difficult to achieve than control of those inhabiting the soil. The principle of controlling this group of pests is the direct opposite of that for controlling soil-inhabiting pests in the following ways:

(1) **Initial contact** of the insecticide with the insect;
(2) **Future contact** of the insect with insecticide residue left on the thatch, by the insect feeding on treated stems or foliage, sucking fluids from the plant, or a combination of both of these.

Initial Contact. Liquid applications have the greatest impact on the insect population within the first 24 to 48 hours after application. Insects are either killed by **initial contact** with the insecticide the day application is made, or the night following treatment by **contact with residues** and **consumption of treated stems or foliage**. The latter is particularly important for control of chewing insects such as cutworms, sod webworms, armyworms, and etc.

The initial impact of granular treatments is less than that from liquids because the insecticide is applied dry. Contact begins only when the granule absorbs moisture and releases the insecticide. For this reason, granular formulations are not effective against pests such as aphids or mites unless the insecticide is systemic.

Future Contact. The residual activity of liquids and granules have their respective advantages and limitations in terms of control beyond the first 24 to 48 hours after application.

In addition to providing initial control of leaf and stem pests, **liquids** leave residues that remain in the thatch, providing control of thatch inhabitants for some time. The length of this residual activity is, however, generally shorter than that of granules. The reasons for the longer residual with **granules** are: they are deposited in the thatch where, unlike liquids on the foliage, they are protected from light which may break down the insecticide; also, insecticide leaches off granules over a period of time which usually exceeds the length of residual activity from liquids.

photo: Gandy Co.

Drop spreaders usually provide more precise distribution of granular formulations, whether used on lawns or larger turf areas. However, there is always the problem of avoiding overlaps and gaps between runs. Insecticide granules generally must be irrigated in order to cause release of the active ingredient.

Generally, the length of residual activity varies considerably with the insecticide and environmental conditions under which it is used. Data expressing residual as parts per million (ppm) over time have been obtained for most insecticides. However, the practical meaning of these data, in terms of actual impact on insect populations in turf, is not understood, particularly where the influence of diminishing residues on insects that survive the initial impact of the treatment is concerned. Data taken from 17 Ohio home lawns treated by a lawn care firm show these diminishing residues were important in removing chinch bugs that survived the treatment and those that hatched or migrated in 10 to 14 days after treatment.

Systemic Insecticides. Systemic insecticides are materials which are sufficiently water soluble to be absorbed by the plant roots and *translocated* to the crown, stems and leaves. This process may take from a few days to a few weeks, depending on temperature, moisture and plant growth. Some materials may also be absorbed by and remain in the leaves (*translaminar absorption*).

Billbug larva feeding in stem is a target for translocated insecticides.

Piercing-sucking mouthparts of the hairy chinch bug make it a target for translocated & translaminar insecticides.

The degree, amount or extent of systemicity varies with the insecticide. The key to the extent to which systemicity controls turfgrass insect pests is the concentration of the active ingredient in the fluids of the plant part being fed upon, and susceptibility of the various stages of the pest to that concentration. Some insecticide products claim, and in fact, are systemic, but the level of systemicity may have little, if any, controlling effect on the target pest(s). Product labels usually do not mention which pests are controlled by the systemic properties of the material. Further, *insecticides are often not investigated thoroughly enough to learn the spectrum of pest insects affected* by the systemic properties of the material.

Systemicity usually provides little or no control of soil-inhabiting pests such as grubs, but can be very effective in controlling insects that feed on the crown, stem and leaf. The earliest developmental stages of such insects are most susceptible.

The recent labeling of chloronicotinyl and thionicotinyl insecticides has forced turf managers to rethink their usage strategies. Both classes of insecticides provide long residual control of soil-inhabiting insects, and they also have systemic properties which show high potential for suppressing, if not controlling, insect pests that inhabit the crown, stems and leaves. These attributes have *renewed interest in the role of these properties to expand the spectrum of pests controlled by these new classes of insecticides*. Expect further labeling of insecticides in these classes.

Sprays made using "shower droplet" style nozzles usually apply larger volumes and the larger droplets tend to better penetrate the canopy.

The Combined Effect. Contact with and ingestion of plant parts or fluids, including thatch, are the primary means by which insect pests that inhabit turf leaves, stems, crowns, and thatch are controlled with insecticides. *It is the collective impact of these modes of action that provide control*. When considering the purchase of a product, **WE STRONGLY RECOMMEND** that the consumer inquire and become knowledgeable about the products' mode(s) of action and whether these mode(s) actually relate to controlling target pest(s).

Application Objectives. The objective of *liquid application* is to deposit insecticide on the foliage and stems or in the crown and thatch to a depth frequented by the target pest (=*the TARGET ZONE*). The volume of liquid applied must be adequate for complete and uniform coverage. Coarse sprays are best because they produce minimal drift. *When the thatch or surface is very dry, irrigation the day before treatment helps facilitate movement of the liquid into the thatch*.

The spray volume needed is also dependent upon the *zone of activity* and *location of the target pest*, as well as the nature and density of the thatch. Thatch in southern turfgrasses, such as St. Augustinegrass, is denser and deeper than that of northern turfgrasses. In order to penetrate the deeper thatch, a high spray volume may be needed. On northern turfgrasses, thatch is usually one inch or less. Under these conditions, 2 to 4 gallons per 1000 ft^2 is usually adequate. When there is little or no thatch, 1 to 2 gallons may be sufficient. Cutworm control of golf course greens is usually accomplished with one gallon per 1000 ft^2.

Some Words of Caution. Reducing the volume of liquid sprays may be more convenient and/or profitable, however, it usually leads to reduced levels of control and/or the need for additional applications. *Carefully consider the OBJECTIVE of the application and THE TARGET PRINCIPLE before reducing the volume of spray applied.*

The initial objective of granular application is to get the material as deeply into the thatch as possible. To do so, the foliage should be dry so the granules fall off the grass blades easily.

Post-treatment Irrigation. The advisability of irrigation following liquid application of a contact insecticide depends on the target pest. If the insect is one that *lives on* the foliage and stems (e.g., bermudagrass mite, greenbug), or one that consumes the

Low volume sprayers require careful attention to irrigation to move the control product to the TARGET ZONE.

Irrigation before an application can moisten thatch and irrigation after an application will also assist movement of a control product into the thatch if that is where the target resides.

foliage (e.g., cutworms, armyworms, sod webworms) or crowns (e.g., billbug larvae), irrigation and mowing should be delayed for at least 24 hours or as long as practical after application. This allows time for contact with or consumption of treated foliage, stems or crowns.

If the primary target is a **thatch inhabitant** (e.g., chinch bug, adult billbug) and the volume applied is 1 to 2 gallons per 1000 ft² or less, a light irrigation is sometimes helpful if done **before the spray dries**. When large volumes of spray are applied, post-treatment irrigation is not necessary. Rainfall or irrigation a week or so after treatment can briefly reactivate the insecticide residue in the thatch.

Irrigation following application of **granular insecticides** is essential to move the insecticide off the granules and into the thatch; therefore, only a light irrigation (1/6 inch or so) is needed. Excess irrigation is not helpful.

Liquid or granular formulations of **systemic insecticides** requiring root uptake should be irrigated after application to hasten translocation to the crown, stems and leaves.

Biological Control

Insect Parasitic Nematodes. Commercial preparations of the parasitic nematode, **Steinernema** spp., have been shown to be effective in controlling bluegrass billbug and annual bluegrass weevil, as well as, leaf eating caterpillars such as cutworms, armyworms and sod webworms. **Close attention to label recommendations**, especially pre- and post-treatment irrigation and soil moisture, is essential for success because nematodes are highly susceptible to desiccation and UV light.

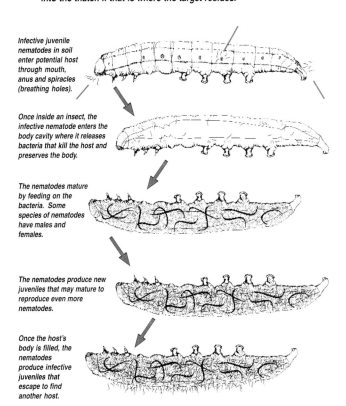

Infective juvenile nematodes in soil enter potential host through mouth, anus and spiracles (breathing holes).

Once inside an insect, the infective nematode enters the body cavity where it releases bacteria that kill the host and preserves the body.

The nematodes mature by feeding on the bacteria. Some species of nematodes have males and females.

The nematodes produce new juveniles that may mature to reproduce even more nematodes.

Once the host's body is filled, the nematodes produce infective juveniles that escape to find another host.

Diagram of typical insect parasitic nematode life cycle.

With billbug and annual bluegrass weevil larvae, the **principle of control** is to apply nematodes when larvae are present. For control of leaf eating caterpillars, the nematodes are most effective when a regular application

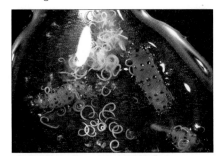

Sod webworm larva infected with Steinernema nematodes. Large ones are adults and small ones are juveniles.

program is followed beginning when the first eggs begin hatching and continuing at 14 to 21 day intervals. **The TARGET is early-stage larvae, before visible damage occurs.**

BT. Products containing the microbial toxin derived from the bacterium **Bacillus thuringiensis** (BT) adequately control thatch-inhabiting caterpillars (especially sod webworms and armyworms) if enough of the toxicant is ingested by young larvae during the first to third instars of development. Irrigation should be omitted since the principle of control is ingestion of the treated turf.

Chinch bugs are commonly infected by Beauveria (three on right) (uninfected chinch bug adult on left).

Beauveria. Formulations of spores from the naturally occurring insect parasitic fungus, *Beauveria bassiana*, are registered for use on turfgrass. Though most, if not all, crown, thatch, stem and leaf inhabiting insects are susceptible to this fungus as it naturally occurs in turf, commercial formulations of the spores have had limited success. The **principle of control** is that the spores adhere to the exoskeleton of the insect and germinate. The fungal hyphae penetrate the body, and once inside, the mycelia completely fill the body of the insect, causing death.

Bluegrass billbug adults are infected and killed by Beauveria.

Cultural Control

Resistant Varieties. Other than the endophytic varieties (which may also be considered "resistant"), few varieties of cool-season turfgrass species have been evaluated for insect resistance or are commercially marketed as being insect resistant. Several varieties of warm-season grasses resistant to certain pests are available. Flora Tex® bermudagrass has been shown to be resistant to the bermudagrass mite, and Cavalier® zoysiagrass is resistant to the fall army-worm and tropical sod webworm.

While the use of resistant varieties is useful for suppression, if not complete control of turfgrass insect pests, the ability of insects to adapt (i.e., overcome the resistance) to these varieties places limitations on even this approach to control. Insects having multiple generations each year (greenbug aphid, southern chinch bug in south Florida) are most capable of such adaptation. The southern chinch bug has developed the ability to feed on the resistant St. Augustine varieties Foratam® and Floralawn® in some areas of Florida and south Texas.

Variable Susceptibility. Field studies conducted in Nebraska have shown that thinner stemmed varieties of Kentucky bluegrass are less prone to damage from bluegrass billbug than are varieties with thicker stems. Apparently, the thinner stemmed

varieties are less acceptable to the insect for egg laying. Research in New Jersey showed hard and Chewings fescues were less prone to damage by the bluegrass billbug while some strong creeping red fescues had considerable damage. Field and laboratory evaluations in Maryland have shown considerable variation in chinch bug damage among varieties of fine fescue and Kentucky bluegrass.

While research has demonstrated that currently available varieties of turfgrasses vary in their susceptibility to damage from crown, thatch, leaf and stem inhabiting insects, the names of these varieties are usually not publicized for various reasons, but can be discovered only by a thorough search of the scientific literature. However, we offer the following table for the reader's consideration.

Resistance of Kentucky Bluegrass Cultivars to Billbugs[a]

Resistant	Eagleton™, Eclipse™, Washington™, Wabash™, America™, Adelphi™, Unique™, Fylking™, Kenblue™, (common)
Moderately Resistant	Midnight™
Highly Susceptible	Broadway™, Parade™, Cheri™, Sydsport™, Columbia™

[a] From Jennifer M. Johnson-Cicalise, Rutgers University, 1997 *Turfgrass TRENDS*™

Endophytic Grasses. Varieties of perennial ryegrass (*Lolium* spp.), fine and turf-type tall fescues (*Festuca* spp.) containing a fungus that produces toxins (a variety of alkaloids) are available for suppressing development of damaging populations of chinch bugs, sod webworms, armyworms, fall armyworm, greenbug aphid and billbugs. Research in Ohio has shown that initial seeding for establishment or overseeding into existing turf can be effective. Control or suppression is achieved by the direct effects of the endophyte's toxins on the insect when above ground plant parts are ingested. The toxins cause disruption of normal feeding and behavior which may render the insect more susceptible to natural enemies.

Limitations. While it is known that endophytes produce toxins (a complex of alkaloids and alkaloid groups), their individual and collective concentrations within the plant may vary with growing conditions, fertility levels and management practices. Moreover, little is apparently known about the concentration of the alkaloids and/or alkaloid groups needed to affect suppression of the various crown, thatch, leaf and stem-inhabiting insects. Despite the impact which the endophytes have on leaf, stem, thatch and crown-inhabiting insects, this impact is secondary, at best, to the total agronomic acceptability of the grass when seed selections are made.

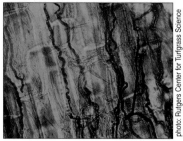

Microscopic view of endophyte fungal strands (dark in picture) located between leaf cells.

Patches of this St. Augustinegrass lawn were damaged by southern chinch bug. Note the pattern of damage caused by blocked irrigation heads which allowed the chinch bugs to escape infection from Beauveria, the white fungus of insects.

Cultural Practices

Regular irrigation or rainfall can have a major suppressing effect on the development of damaging populations of many crown, thatch, stem and leaf-inhabiting insect pests. **Beauveria bassiana, the white fungus that infects most of these pest, is especially infections under moist conditions.** Where possible, the use of fungicides which kill this white fungus should be avoided.

Any cultural practice, including irrigation, fertilization and mowing that produces and maintains a vigorous plant capable of tolerating or outgrowing insect damage is the right thing to do. However, except for regular irrigation or rain optimizing *Beauveria* infection, these practices will not necessarily prevent pest infestations. A vigorous, well irrigated turf is an attractive egg laying site for grub producing beetles such as Japanese beetle and masked chafers. Never the less, it is the right thing to do.

Thatch is both beneficial and detrimental to good turfgrass management. Insects such as chinch bugs, billbugs, sod webworms,

Core aerification is a common practice used to help manage thatch in turf. Cores must be deep enough to pull up soil below the thatch layer which aids in decomposition of thatch and provide aerification. However, on golf greens, cores are usually removed.

cutworms and others live and feed in thatch when it is present. Therefore, management practices like top dressing, verticutting, or core aerification for removal of excessive thatch, helps reduce infestations of these pests.

Pest insect control is also a cultural practice. Judicious, conservative use and careful selection of insecticides and/or biological controls that are effective in controlling Target Pest(s) and have minimal effect of the environment and natural control organisms is also the right thing to do.

Natural Control

While the turfgrass environment harbors the insect and mite pests mentioned in this book, it also harbors **many organisms** which exert a constant suppressing (=controlling) influence on these pests. Without that influence, pest populations would grow unabated. Among the organisms are: surface and soil inhabiting bacteria, nematodes and fungi (e.g., *Bacillus popilliae*, *Steinernema*, and *Beauveria*) that cause **diseases**; ground beetles, rove beetles, ants, bigeyed bugs, mites and more that are **predators**; and many species of tiny wasps and flies that **parasitize** insect eggs, larvae, pupae and adults.

Unavoidably, necessary application of certain insecticides for control of crown, thatch, stem and leaf inhabiting pests can (though temporary) reduce some of these natural controls. Judicious use and careful selection of insecticides and fungicides that are effective on controlling Target Pests and have minimal effect on Natural Control Agents, helps conserve the beneficial organisms.

The recent development and registration of the chloronicotinyl class of insecticides which are systemic in the turf plant and primarily toxic to insects that feed on the plant is **GOOD NEWS FOR CONSERVATION OF NATURAL CONTROLS**. Most predators and parasites which do not feed on the plant are not killed. Further, **this class of insecticides apparently modifies the behavior of the Target Pests to the extent that they loose their ability to defend against many natural enemies.**

Ground beetle adults commonly feed on insect eggs and larvae.

This Tiphia wasp seeks out grubs to paralyze and use as a larval host. However, the adults (wasps) are not "appreciated" by the general public.

Bigeyed bugs are general predators. This adult is feeding on another bigeyed bug nymph!

87

Notes

Chapter 8

Control Approaches and Programs

So far in this book, we have provided:

(1) **our perspectives** on PREVENTIVE, CURATIVE, and TOLERANCE approaches to controlling destructive turf insects;

(2) **many color photos** of these insects as well as information on life cycles, habits and diagnosis;

(3) **principles of control**, including chemical, biological, cultural and natural approaches.

The factors involved in deciding which approach(es) is taken were: PERSPECTIVES of the person(s) making the decision(s); FINANCIAL CONSIDERATIONS; TURF QUALITY STANDARDS; and PEST SPECTRUM.

In this chapter we propose:

(1) development of a **PEST SPECTRUM AND TARGET CALENDAR**;

(2) **INTEGRATION OF APPROACHES INTO PROGRAMS** for Control of **PRIMARY** and **SECONDARY PESTS**;

(3) **EVALUATION** of the APPROACHES and PROGRAMS you use in relation to your PEST SPECTRUM AND TARGET CALENDAR.

Developing a Pest Spectrum and Target Calendar

While the first three factors in deciding which approach(es) should be taken are always part of the process, the Pest Spectrum is often not considered. The Principle of the **MULTIPLE TARGET CONCEPT** is: "**insect pests of turfgrasses rarely occur one at a time at any one time**." Regardless of the approach or combination of approaches (chemical, biological, cultural or natural), the impact on not only the primary target pest, but the other pests and their life stages throughout the growing season, should be considered.

For Example - The **Primary Target** of a preventive approach using the insecticide imidachloprid, applied in May, might be grubs to prevent damage in late summer or fall. While prevention of grubs is the **intended target** of the treatment, **other pests (= Secondary Targets)** present at the time of the application and over the active residual life of the treatment, **may also be controlled or suppressed**. The same consideration would apply for any month that imidachloprid or any other treatment would be applied.

In northern lawns, white grubs may be the PRIMARY TARGET. During the season, bluegrass billbugs (upper left), sod webworms (upper right), hairy chinch bugs (lower right), or greenbug aphids (lower left) may also be present, and are thus, potential - SECONDARY TARGETS. In other lawns or other regions, any of these secondary targets may be the primary target!

Assembling **Your** Pest Spectrum and Target Calendar

The following section of this chapter provides the reader with an opportunity to **reevaluate**, **review**, and **refocus** on control of damage from destructive turf insects and mites by assembling a **Pest Spectrum and Target Calendar specific to _your_ location**. The required inputs for this "exercise" are:

(1) Your knowledge, experience and the history of pest occurrences in the area.
(2) The knowledge and experience(s) of other turfgrass managers, and,
(3) The previous chapters of this book.

General Pest Spectrum and Target Calendars for golf courses and lawns in northern (page 91) and southern turf (page 93) are provided for your reference. **Copy the blank form** provided on the following pages, and, by using the footnotes in the general spectrums provided, **assemble a pest spectrum and target calendar for _your_ location**.

Starting with the month before the growing season begins, list the adult, larval or nymphal stages of all pests known to occur for each month (or two-week periods if activity times are shorter than a month) of the entire growing season at your location. We think this "exercise" will enhance the value of this chapter on approaches and development of Preventive/Curative Control Programs.

The PRIMARY TARGET of most southern golf courses is mole crickets. However, bermuda-grass mites (upper left), and/or white grubs (upper right), and/or hunting billbug (lower right), and/or fall armyworm (lower left) (=SECONDARY TARGETS) may be present at the time of treatment. Assembly of YOUR <u>Pest Spectrum and Target Calendar</u> requires knowing when and where all of these pests occur during the season.

Notes
Some questions you should consider when assembling <u>your</u> Pest Spectrum and Target Calendar.

What insect and mite pests, and/or symptoms of their damage have been seen?

What other insects and mites are known to infest the turf at your location(s) (this book should help!)?

What are the different stages of each pest and **when** do they occur in your area?

Which controls (chemical, biological or cultural) will you select? <u>What is the spectrum of their efficacy</u> against the pests you identified in <u>your</u> Pest Spectrum and Target Calendar, and <u>when</u> is the optimal time to use them?

General Pest Spectrum and Target Calendar
for Northern Golf Courses and Home Lawns

Northern Golf Course
Pest Spectrum and Target Calendar

April	May	June	July	August	September	October
• Grubs (ow)	• Grubs (ow)		• Grubs	• Grubs	• Grubs	• Grubs
• Grubs (prv) (s/f)	• Grubs (prv) (s/f)	• Grubs (prv)	• Grubs (prv)			
• Green JB/l (ow)	• Green JB/l (ow)	• Green JB/a	• Green JB/a-l	• Green JB/l	• Green JB/l	
• BTA/a	• BTA/a	• BTA/l	• BTA/l	• BTA/a	• BTA/l-a	• BTA/a
• Billbug/a (ow)	• Billbug/a (ow)	• Billbug/l	• Billbug/l	• Billbug/a	• Billbug/a	• Billbug/a
• Billbug/l (prv) (sp/s)	• Billbug/l (prv) (sp/s)					
• SWW/l	• SWW/l	• SWW/l	• SWW/l	• SWW/l	• SWW/l	• SWW/l
	• Cutworm	• Cutworm	• Cutworm	• Cutworm	• Cutworm	
• Cutworm (prv) (sp)	• Cutworm (sup) (sp)	• Cutworm (prv/sup)			• Fall AW	• Fall AW
• Ant (sup) (sp/s)	• Ant (sup)	• Ant (sup)	• Ant (sup)	• Ant (sup)	• Ant (sup)	• Ant (sup)
• Greenbug (prv) (s)	• Greenbug (prv) (s)	• Greenbug	• Greenbug	• Greenbug	• Greenbug	
• ABGW/a (prv) (sp)	• ABGW/a/l (prv) (sp)	• ABGW/l/a	• ABGW/l/a	• ABGW/a/l		
	• Frit Fly/a/l (prv) (sp/s)	• Frit Fly/l/a	• Frit Fly/a/l (prv) (s/f)	• Frit Fly/l/a	• Frit Fly/a/l (prv)	• Frit Fly/a
• WinterGM	• WinterGM					

Grubs	= Annual grub species	**Fall AW**	= Fall armyworm	a	= Adult target	
BTA	= Black turfgrass ataenius	**Ant**	= Turfgrass ant	l	= Larval target	
Billbug	= Bluegrass billbug	**Greenbug**	= Greenbug aphid	n	= Nymphal target	
SWW	= Sod webworms	**ABGW**	= Annual bluegrass weevil	ow	= Overwintered stage	
Cutworm	= Black cutworm	**Frit Fly**	= Frit Fly	sp	= Spring	
WinterGM	= Winter grain mite	**GreenJB**	= Green June beetle	s	= Summer	
				f	= Fall	
prv	= prevention	sup	= suppression			

Northern Lawns
Pest Spectrum and Target Calendar

April	May	June	July	August	September	October
• Grubs (ow)	• Grubs (ow)		• Grubs	• Grubs	• Grubs	• Grubs
• Grubs (prv) (s/f)	• Grubs (prv) (s/f)	• Grubs (prv)	• Grubs (prv)			
• Green JB/l (ow)	• Green JB/l (ow)	• Green JB/a	• Green JB/a-l	• Green JB/l	• Green JB/l	
• Chinch Bug/a	• Chinch Bug/a-n	• Chinch Bug/n	• Chinch Bug/a-n	• Chinch Bug/a-n	• Chinch Bug/a-n	• Chinch Bug/a
• Chinch Bug (prv) (s)	• Chinch Bug (prv) (s/f)	• Chinch Bug (prv) (s/f)	• Chinch Bug (prv) (f)			
• Billbug/a (ow)	• Billbug/a (ow)	• Billbug/l	• Billbug/l	• Billbug/a	• Billbug/a	• Billbug/a
• Billbug/l (prv) (sp/s)	• Billbug/l (prv) (sp/s)					
• SWW/l	• SWW/l	• SWW/l	• SWW/l	• SWW/l	• SWW/l	• SWW/l
• Greenbug (prv) (s)	• Greenbug (prv) (s)	• Greenbug	• Greenbug	• Greenbug	• Greenbug	
	• Armyworm	• Armyworm	• Armyworm	• Armyworm		
• Cutworms	• Cutworms	• Cutworms				
• Clover Mite	• Clover Mite					
• WinterGM	• WinterGM					

Grubs	= Annual grub species	**Chinch Bug**	= Hairy chinch bug	a	= Adult target	
Green JB	= Green June beetle	**Ant**	= Turfgrass ant & other ants	l	= Larval target	
Billbug	= Bluegrass billbug	**Greenbug**	= Greenbug aphid	n	= Nymphal target	
SWW	= Sod webworms	**WinterGM**	= Winter grain mite	ow	= Overwintered stage	
Cutworms	= Bronze, variegated, black cutworm	**Clover Mite**	= Clover mite	sp	= Spring	
Armyworm	= (common) Armyworm	**Fleas**	= Fleas and chiggers	s	= Summer	
				f	= Fall	
prv	= prevention	sup	= suppression			

Northern
Lawn / Golf / Sports Turf

Pest Spectrum and Target Calendar

Jan - Feb	March	April	May	June	July	August	September	October	Nov - Dec

Northern Turf Pests

Grubs	=	Annual grub species
BTA	=	Black turfgrass ataenius
Green JB	=	Green June beetle
ABGW	=	Annual bluegrass weevil
Billbug	=	Bluegrass billbug
Chinch Bug	=	Harry chinch bug
Ant	=	Turfgrass ant & other ants
Fleas	=	Fleas and chiggers

Cutworm	=	Black, bronze, variegated cutworm
SWW	=	Sod webworms
Armyworm	=	(common) Armyworm
Fall AW	=	Fall armyworm
Greenbug	=	Greenbug aphid
Frit Fly	=	Frit fly
Winter GM	=	Winter grain mite
Clover Mite	=	Clover mite

a	=	Adult target
l	=	Larval target
n	=	Nymphal target
p	=	Pupa (non-target)
ow	=	Overwintered stage
sp	=	Spring
s	=	Summer
f	=	Fall
prv	=	Preventive application

General Pest Spectrum and Target Calendar
for Southern Golf Courses and Home Lawns

Southern Golf Course
Pest Spectrum and Target Calendar

March	April	May	June	July	Aug	Sept	Oct	Nov	Dec
MoleC/N-a (ow)	MoleC/a (ow)	MoleC/a-n (ow)	MoleC/n	MoleC/n	MoleC/N	MoleC/N-a	MoleC/N-a	MoleC/N-a	MoleC/N-a
MoleC (prv)	MoleC (prv)	MoleC (prv)							
Grubs/l (ow)	Grubs/l (ow)	Grubs/p-a (ow)	Grubs/p-a (ow)	Grubs/l	Grubs/l	Grubs/l	Grubs/l	Grubs/l	
	Grubs (prv)	Grubs (prv)	Grubs (prv)						
Green JB/l (ow)	Green JB/l (ow)	Green JB/l	Green JB/p	Green JB/a	Green JB/a-l	Green JB/l	Green JB/l	Green JB/l	
	Cutworm	Cutworm	Cutworm	Cutworm	Cutworm	Cutworm	Cutworm		
	Fall AW	Fall AW	Fall AW	Fall AW	Fall AW	Fall AW	Fall AW	Fall AW	
	Trop SWW	Trop SWW	Trop SWW	Trop SWW	Trop SWW	Trop SWW	Trop SWW	Trop SWW	
SWW/l (ow)	SWW/l (ow)	SWW/l (ow)	SWW	SWW	SWW	SWW	SWW	SWW/l	SWW/l
Fire Ant	Fire Ant	Fire Ant	Fire Ant	Fire Ant	Fire Ant	Fire Ant	Fire Ant	Fire Ant	Fire Ant
Ants	Ants	Ants	Ants	Ants	Ants	Ants	Ants	Ants	Ants
H Billbug/a	H Billbug/a	H Billbug/a-l	H Billbug/a-l	H Billbug/l	H Billbug/l	H Billbug/l	H Billbug/l	H Billbug/l	H Billbug/l
Bermuda M (dormant turf)	Bermuda M (dormant turf)	Bermuda M/a-n (spring growth)	Bermuda M/n	Bermuda M/a-n	Bermuda M/a-n (damage)	Bermuda M/a (damage)	Bermuda M (damage)	Bermuda M (dormant turf)	Bermuda M (dormant turf)
B Scale (dormant turf)	B Scale (dormant turf)	B Scale (delay greenup)	B Scale	B Scale	B Scale	B Scale (damage)	B Scale (damage)	B Scale (dormant turf)	B Scale (dormant turf)

Mole C	= Mole crickets	Trop SWW	= Tropical sod webworm	a	= Adult target
Grubs	= Annual grub species	Fire Ant	= Red Imported Fire Ant	l	= Larval target
GreenJB	= Green June beetle	H Billbug	= Hunting billbug	n	= Nymphal target
Cutworm	= Black cutworm	Bermuda M	= Bermudagrass mite	N	= Large nymphal target
Fall AW	= Fall armyworm	B. Scale	= Bermudagrass scale	p	= Pupa (non-target)
SWW	= Sod webworm	Ants	= Non-fire ants	ow	= Overwintered stage
prv	= Preventive application	sup	= Suppression		

Southern Lawn
Pest Spectrum and Target Calendar

March	April	May	June	July	Aug	Sept	Oct	Nov	Dec
MoleC/N-a (ow)	MoleC/N-a (ow)	MoleC/a (ow)	MoleC/n	MoleC/n	MoleC/N	MoleC/N-a	MoleC/N-a	MoleC/N-a	MoleC/N-a
MoleC (prv)	MoleC (prv)	MoleC (prv)							
Grubs/l (ow)	Grubs/l (ow)	Grubs/l (ow)	Grubs/p	Grubs/l	Grubs/l	Grubs/l	Grubs/l	Grubs/l	
	Grubs (prv)	Grubs (prv)							
Green JB/l (ow)	Green JB/l (ow)	Green JB/l	Green JB/p	Green JB/a	Green JB/a-l	Green JB/l	Green JB/l	Green JB/l	
SChinch B/a	SChinch B/a-n	SChinch B/a-n	SChinch B/a-n	SChinch B/a-n	SChinch B/a-n	SChinch B/a-n	SChinch B/a-n	SChinch B/a-n	SChinch B/a
	Fall AW	Fall AW	Fall AW	Fall AW	Fall AW	Fall AW	Fall AW	Fall AW	
	Trop SWW	Trop SWW	Trop SWW	Trop SWW	Trop SWW	Trop SWW	Trop SWW	Trop SWW	
Fire Ant	Fire Ant	Fire Ant	Fire Ant	Fire Ant	Fire Ant	Fire Ant	Fire Ant	Fire Ant	Fire Ant
H Billbug/a	H Billbug/a	H Billbug/a-l	H Billbug/a-l	H Billbug/l	H Billbug/l	H Billbug/l	H Billbug/l	H Billbug/l	H Billbug/l
Spittlebug/e	Spittlebug/e-n	Spittlebug/n	Spittlebug/n-a	Spittlebug/a	Spittlebug/a	Spittlebug/a	Spittlebug/a-e	Spittlebug/a-e	Spittlebug/e
Bermuda M (dormant turf)	Bermuda M (dormant turf)	Bermuda M/a-n (spring growth)	Bermuda M/n	Bermuda M/a-n	Bermuda M/a-n (damage)	Bermuda M/a (damage)	Bermuda M (damage)	Bermuda M (dormant turf)	Bermuda M (dormant turf)
B Scale (dormant turf)	B Scale (dormant turf)	B Scale (delay greenup)	B Scale	B Scale	B Scale	B Scale (damage)	B Scale (damage)	B Scale (dormant turf)	B Scale (dormant turf)
		Flea	Flea	Flea	Flea	Flea	Flea	Flea	Flea

Mole C	= Mole crickets	Fire Ant	= Red Imported Fire Ant	a	= Adult target
Grubs	= Annual grub species	H Billbug	= Hunting billbug	l	= Larval target
GreenJB	= Green June beetle	Spittlebug	= Twolined spittlebug	n	= Nymphal target
SChinch B	= Southern chinch bug	Bermuda M	= Bermudagrass mite	N	= Large nymphal target
Fall AW	= Fall armyworm	B. Scale	= Bermudagrass scale	p	= Pupa (non-target)
Trop SWW	= Tropical sod webworm	Fleas	= Fleas & Chiggers	ow	= Overwintered stage
prv	= Preventive application	sup	= Suppression		

Southern
Lawn / Golf / Sports Turf

Pest Spectrum and Target Calendar

Jan - Feb	March	April	May	June	July	August	September	October	Nov - Dec

Southern Turf Pests

Mole C	=	Mole crickets
Grubs	=	Annual grub species
Green JB	=	Green June beetle
Billbug	=	Hunting billbug
Fire ant	=	Red imported fire ant
Chinch Bug	=	Southern chinch bug
Fleas	=	Fleas & chiggers
Ants	=	Non-fire ants

Cutworm	=	Black cutworm
Trop SWW	=	Tropical sod webworm
Fall AW	=	Fall armyworm
Armyworm	=	(common) Armyworm
Clover Mite	=	Clover mite
Bermuda M	=	Bermudagrass mite
B Scale	=	Bermudagrass scale
SWW	=	Sod webworm

a	=	Adult target
l	=	Larval target
n	=	Nymphal target
p	=	Pupa (non-target)
ow	=	Overwintered stage
prv	=	Preventive application

Annual grassy weeds such as annual bluegrass *(above)* and crabgrass can be controlled preventively and curatively. Prevention, in the form of preemergent herbicide, is an easy and effective approach!

Most plant pathologists agree that preventive application of fungicide is the ONLY way to control many turf diseases.

Bluegrass billbug damage *(above)* can be prevented by controlling the adults before they lay eggs. The larvae can be controlled (curatively), if they are discovered early, but prevention is usually surer and more effective approach!

Preventive and Curative Approaches and Programs

The PREVENTIVE and CURATIVE programs for golf courses and lawns outlined in this chapter are presented with the view that one pest (i.e., chinch bug, billbug, etc.) or group of pests (i.e., grubs, cutworms, mole crickets, etc.) is often the **primary** (but not necessarily the only) **concern-focus** of treatments directed by the supervising turfgrass manager or lawn service provider. We call this the "**Primary Target**." The impact of the programs and treatments on **other pests** that also occur at the time of "application" (= **Secondary Targets**) is also considered in this chapter. The programs presented can be modified easily for sports fields.

PREVENTIVE PROGRAMS
FOR NORTHERN TURF

In our view, an insecticide or other form of insect control should be applied only when its use is justified. The major justification for following a **preventive approach** program should be a **past history** of infestation and/or damage and **confidence** that damage from insect pests will reoccur. Such history is based on previous years' experience(s), observations, monitoring and knowledge of the seasonal spectrum of pests occurring at any specific location. Consideration of the impact of a treatment on not only the **Primary Target**, but the **Spectrum of Secondary Pests** as well, **can reduce the number of applications needed to achieve the objective**.

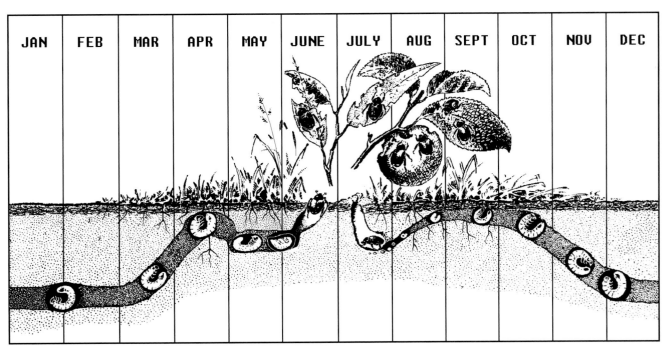

JAN	FEB	MAR	APR	MAY	JUNE	JULY	AUG	SEPT	OCT	NOV	DEC

Prevention of grub damage can be accomplished with application of an insecticide from early May through mid-July. However, application during the first week of May controls a broader spectrum of PRIMARY and SECONDARY pests.

PREVENTIVE PROGRAMS FOR NORTHERN GOLF COURSES

Primary Target - GRUBS

If grubs (black turfgrass ataenius, Aphodius, Japanese beetle, masked chafer, European chafer, Asiatic garden beetle, Oriental beetle, etc.) are determined to be the **Primary Target** and a **Preventive Program** is selected, **early May is the optimal time to apply imidacloprid**. In addition to providing season-long control of these grubs, other secondary pests in the spectrum (billbug larvae, first generation cutworm larvae, and probably greenbug aphids and frit fly) will also be prevented. In our opinion, ants (i.e., *Lasius neoniger*) will also be suppressed. Label directions regarding rate(s) to apply should be followed carefully.

While application of imidacloprid in **June** provides season-long control of grubs, it is too late to prevent the first generation of cutworm and probably billbug larvae as well. Billbug is a significant golf course pest on tee and bunker banks as well as in roughs. Application from **July to mid-August** (according to many State, Cooperative Extension "recommendations") will prevent annual

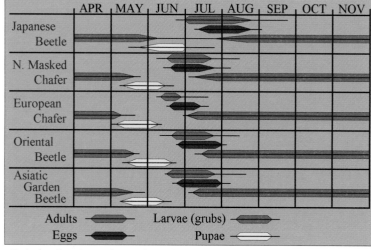

grubs, including green June beetle, but is too late to control most other secondary pests in the spectrum.

In our experience, **preventive control** of Japanese beetle, masked chafer and black turfgrass ataenius grubs **has been achieved with a single application of the new thianicotinyl, thiamethoxam in May or June or July**. Larvae of billbugs, cutworms and

White grubs in the soil-thatch interface, may be the PRIMARY TARGET but not the ONLY TARGET.

sod webworms existing during and for a time after the time of application are also controlled. The complete activity spectrum of this new insecticide is still being explored at the time of this writing.

Season-long preventive control of black turfgrass ataenius and *Aphodius*, Japanese beetle and masked chafer grubs can be **achieved with June applications of halofenozide**. Control of European chafer and Asiatic garden beetle is limited. Infestations of billbugs, cutworm and sod webworm larvae existing at the time of application may also be controlled with treatment at this time.

Application of halofenozide from **July thru early August** also prevents infestation of the above grubs and controls existing infestations of cutworm and sod webworm. Treatments applied from **mid-August to mid-September** control Japanese beetle and masked chafer and may provide some degree of control of sod webworm larvae that would normally overwinter.

At the time this book was published, preventive control of green June beetle larvae with halofenozide had not been investigated.

Primary Target - ANNUAL BLUEGRASS WEEVIL

On golf courses where **grubs and annual bluegrass weevil are major targets**, a combination of imidacloprid plus a pyrethroid insecticide applied from **mid- to late-April** prevents damage from first and second generation annual bluegrass weevil larvae. This treatment should also prevent larval infestations of billbug, black turfgrass ataenius, Japanese beetle, masked chafer, European chafer, and (we think) first generation cutworms.

On golf courses where the grub species are not major targets, single application of the labeled pyrethroid insecticides during the **third week of April** should prevent damage from annual bluegrass weevil larvae. **The principle** of this approach is to target overwintered adults as they return to annual bluegrass to begin egg laying. **Timing is critical**. Treatment must be applied before significant egg laying has occurred.

The impact of such a treatment program on the spectrum of secondary target pests occurring at the time of application has not been well studied. However, because

	APR	MAY	JUN	JUL	AUG	SEP	OCT	NOV
Japanese Beetle								
N. Masked Chafer								
European Chafer								
Oriental Beetle								
Asiatic Garden Beetle								

Adults ◄► Larvae (grubs) ◄►
Eggs ◄► Pupae ◄►

Understanding when various grub species lay eggs and early instar grubs begin feeding at the soil-thatch interface determines the best time to make preventive control applications in order to target all species of grubs (=Primary Targets) and secondary targets as well.

Annual bluegrass weevil damage can be prevented.

these insecticides are labeled for and known to be residually toxic to black turfgrass ataenius and billbug adults, larval infestations of these pests <u>should also be prevented</u>.

BTA adults begin egg laying when Vanhoutte spirea comes into bloom. Application of a control product that kills these adults stops egg laying and prevents damage.

Primary Target -
BLACK TURFGRASS ATAENIUS

In situations where **BTA is the <u>only</u> grub of concern**, another preventive option has been used successfully. The <u>principle of control</u> involves application of **chlorpyrifos or a labeled pyrethroid** to target overwintering adults just as egg laying begins. In the northern states, this event coincides with the onset of full bloom of Vanhoutte spirea (*Spirea vanhouttei*), usually **early to mid-May**. <u>The objective</u> of the treatment is to deposit the insecticide into the first 1/4-inch of the thatch so that residues will kill adults as they land on the turf to hide or burrow to lay eggs. Treatments should be **syringed immediately after application** to wash the insecticide off the grass blades into the thatch.

A preventive application of <u>imidacloprid</u> or <u>thiamethoxam</u> during the **<u>first week of May</u>** or <u>halofenozide</u> in **early June** to control other major grub targets, also controls black turfgrass ataenius, *Aphodius* and a spectrum of other pests. These insecticides can also be used successfully when black turfgrass ataenius is the primary target.

Primary Target - BLUEGRASS BILLBUG

The bluegrass billbug is a <u>significant cause of damage to Kentucky bluegrass and non-endophytic perennial ryegrass</u> around golf course greens and sand bunkers, on tee and green banks, in roughs and turf around the club house. Damage can be as subtle as a chronic thinning of the stand. **Symptoms are often misdiagnosed as irrigation not reaching the turf, drought, or**

This thinning and damage spots are caused by bluegrass billbugs that prefer the sunny, dry bunker banks of golf courses. When these areas are treated in May, this damage does not appear!

disease such as dollar spot. If uncontrolled for extended periods, the Kentucky bluegrass portion of a sward continues to diminish over time. Kentucky bluegrass varieties vary in their susceptibility to this pest (pp. 86, Chapter 7).

An <u>application of **imidacloprid** or **thiamethoxam**</u> during the **first week of May** or **halofenozide** in **early June** for prevention of major targets also prevents bluegrass billbug damage and controls a spectrum of other pests.

Endophytic Ryegrasses - Where appropriate and acceptable, **perennial ryegrasses with endophytes are an effective preventive approach** to control of bluegrass billbug. Initial seeding for establishment or overseeding into existing turf is effective. Sod webworms, armyworms, and fall armyworms are included in the spectrum of pests suppressed.

Primary Target - OTHER PEST INSECTS

CUTWORMS

When cutworms (mainly the black cutworm) are the **Primary Target** of concern on golf courses, **a preventive approach is not recommended**. <u>We discourage</u> adding an insecticide to a treatment of which the objective is fertilization, and/or growth regulation, and/or disease control "just in case" there <u>may be</u> cutworms present. Instead, <u>we recommend a curative approach</u> and application of a control (if necessary) when evidence of damage first appears.

Biological Control of Cutworms and Armyworms. Products containing the insect parasitic nematodes, *Steinernema carpocapsae* or *S. feltiae*, <u>are available and can effectively control cutworms, armyworms and sod webworms</u>. Close attention to label directions, especially <u>pre- and post-treatment irrigation</u> and soil moisture requirement, is <u>essential to success</u>.

Black cutworms usually make depressed, "pock marks" in short cut bentgrass. Imidachloprid (and probably thiamethoxam) applied in early May prevents the first episode of such damage.

Sod webworm damage in summer may appear as irregular trails or darker marks on greens. This is often mistaken for disease or other causes.

A program of regularly scheduled applications, beginning when the first eggs begin to hatch and continuing at a 14- to 21-day interval, thereafter, has been shown to prevent damage. *Affected larvae die in their burrows, not on the turf surface.*

SOD WEBWORM

Generally, sod webworms have not been considered a pest worthy of concern on golf courses. Our experience, based upon observations on golf courses and communication with golf course superintendents from Ohio to Nevada to Florida, indicates otherwise!

Sod webworm larvae *commonly overwinter in greens*, practice greens and tees (and many other areas) on golf courses. *The overwintered larvae resume feeding in early spring* (mid-April to early May in Ohio) by constructing a C-shaped cover of webbed-together topdressing over its burrow. The sand cover is just below the mowing level of the turf. The larva feeds on the turf under the cover, which is made larger as more food is required. We have seen greens with 1 to 2 covers per square foot.

In addition to the sand covers being unsightly and possibly interfering with ball roll, *the larvae* under them *are a major reason for the probing of starlings and other birds in*

early spring (April). This is well before the first generation of cutworms occurs.

When necessary, *spring damage can be prevented* by treating the turf areas of concern with an insecticide from *late September to mid-October* to kill the larvae that would otherwise overwinter. *May* application of imidachloprid for preventive control of grubs or other primary targets has not controlled overwintered sod webworms.

TURFGRASS ANT - *Lasius neoniger*

As of the writing of this book, a consistent preventive program for control of this ant had not been developed. Our research has shown that treatments of chlorpyrifos (alone or as a bait formulation) or one of the pyrethroids, such as bifenthrin or deltamethrin, applied as soon as activity begins (*late April to early May*) suppresses mound construction for 90 days or more. Application of imidachloprid, fipronil, or thiamethoxam with a rapid acting insecticide at first ant activity has suppressed early mound construction and subsequent activity for more than 120 days. Spring application of bifenthrin or imidachloprid suppressed ant mounding 50 to 60% the following spring compared to untreated areas.

Overwintered sod webworm larvae commonly make C-shaped marks on golf greens in April and May (left). These spring larvae are difficult to control because they rest in deep burrows (right) and their surface feeding occurs under the cover of webbed together topdressing.

A consistant preventive program for control of the turfgrass ant is expected in the very near future.

Notes

PREVENTIVE PROGRAMS FOR NORTHERN LAWNS

Primary Target - GRUBS

If grubs (Japanese beetle, masked chafer, European chafer, Asiatic garden beetle, or Oriental beetle) are the **primary target**, and a preventive approach is selected by home owners, lawn service client, or lawn service provider, **late April through May is the optimal time to apply imidacloprid**. Label directions regarding rates to apply should be followed carefully. In addition to providing seasonal control of grubs, the systemic activity of the material should also provide season-long control of billbug larvae (a frequent cause of damage), greenbug aphid, and (in our opinion) at least suppression of chinch bug below damaging levels. Little or no control of overwintered sod webworm larvae is expected and influence of succeeding sod webworm generations is unknown.

Mid- to late-June application of imidacloprid provides season-long control of grubs, but probably will not provide adequate control of billbug larvae. The impact of this treatment on the first generation of chinch bug and sod webworm is not known at this time. Application from July to mid-August will prevent annual grubs

(including green June beetle), but is probably too late to impact other secondary pests in the spectrum (though this is actually unknown).

Our experience has shown that **May application of thiamethoxam** provides season-long control of Japanese beetle and masked chafer in Ohio. Studies in other states have shown similar control of European chafer and Oriental beetle. Infestations of billbug and sod webworm larvae can also be controlled by May applications, but no research has been performed to determine the effect of a May application on future generations of sod webworms

June application of halofenozide provides season-long control of Japanese beetle and masked chafer. Control of European chafer and Oriental beetle has been maximized when halofenozide is applied at egg laying. Infestations of billbug and sod webworm larvae existing at the time of application may also be controlled with treatment at this time. Application from July to early August prevents infestation of Japanese beetle and masked chafers and controls existing sod webworm larvae. Treatments applied from mid-August thorough mid-September also control these grubs.

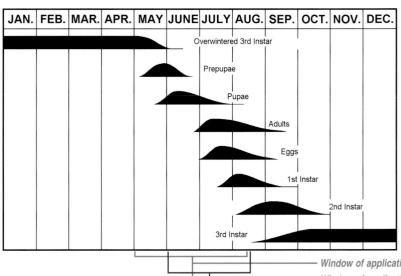

JAN.	FEB.	MAR.	APR.	MAY	JUNE	JULY	AUG.	SEP.	OCT.	NOV.	DEC.

Overwintered 3rd Instar

Prepupae

Pupae

Adults

Eggs

1st Instar

2nd Instar

3rd Instar

— Window of application for imidachloprid or thiamethoxam.
— Window of application for halofenozide.

Japanese beetle life history, typical of central New Jersey to central Ohio.

Application of imidacloprid or thiamethoxam from late April through May will control this annual grub as well as provide season-long control of bluegrass billbug larvae and suppression of chinch bug (and possibly other Secondary Targets).

JAN.	FEB.	MAR.	APR.	MAY	JUNE	JULY	AUG.	SEP.	OCT.	NOV.	DEC.

3rd Instar   (two-year brood)

Prepupae

Pupae

Adults

Eggs

1st Instar

2nd Instar

(two-year brood)

3rd Instar

European chafer life history in Upstate New York.

When compared to Japanese beetle (above), European chafer adults fly earlier and the primary target, first instar grubs, are in the soil-thatch interface 3 to 4 weeks earlier than Japanese beetle first instar grubs. Preventive application in late or early May of imidacloprid, thiamethoxam, or halofenozide has appeared to be somewhat less effective against European chafer than Japanese beetle larvae. Therefore, application at adult flight is the usual suggested time of application for control of European chafer. Our experience indicates that application of imidacloprid or thiamethoxam at maximum label rates in May provides good control of this grub as well as a range of SECONDARY TARGETS (i.e., billbug & chinch bug).

Primary Target - **BILLBUG**

Damage from billbugs is the most commonly misdiagnosed and most common cause of damage to lawns. Damage is often misdiagnosed as drought or sod webworm injury. Differentiating damage from the two pests and drought is easily accomplished (See: Chapter 5, pages 53 and 59).

Damage can be prevented. Diazinon or a labeled pyrethroid applied between ***mid-April and mid-May*** kills overwintered billbug and chinch bug adults and sod webworm larvae. Treatment at this time prevents development of a first generation of these pests (= ***Multiple Primary Targets***). Application after mid-May could result in some billbug damage if significant egg laying occurred before the application.

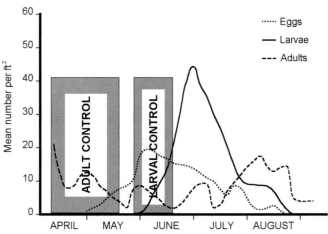

Insecticide application in April into May kills overwintered billbug adults and prevents larval damage in June and July.

Primary Target - **CHINCH BUG**

At locations where chinch bug has one generation each year, a labeled pyrethroid applied between ***mid-April and early-May*** should provide season-long control. In areas with two generations of chinch bug, the need for a second treatment depends upon the abundance of rainfall or irrigation during development of the second generation. Further, if adjacent lawns were not treated for chinch bug, there is a higher risk of chinch bugs migrating into the treated lawn.

When moisture for good turf growth is adequate during late August and September, a fungus disease, *Beauveria*, infects and kills many chinch bugs. In many, if not most cases, the population can be reduced to the point where a second insecticide application is unnecessary. The fact that most northern lawn care programs include a fertilizer application at this time also helps the turf withstand feeding from the second generation. Application of fungicides significantly reduces *Beauveria* effectiveness.

If August or September are ***dry (a condition under which chinch bugs thrive)***, reinfestation limited to the border can be caused by chinch bugs moving from adjacent untreated turf. Some border damage from first and second generation migrants is possible under such circumstances. If the adjacent turf has been treated or has a low level of infestation, reinfestation is very unlikely since the source of reinfestation is not there. In this case, the spring treatment could take care of the chinch bug problem for the year.

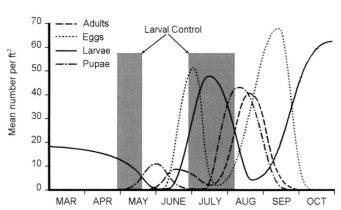

Adult chinch bugs overwinter and become active in April into May. If these adults are controlled, damage in June and July is prevented.

Primary Target - **SOD WEBWORMS**

Sod webworms rarely require a preventive control strategy. However, when drought conditions occur in consecutive years, sod webworms can cause unacceptable damage and a preventive control may be warranted. Some sod webworm species, especially the cranberry girdler, can cause damage in September and October. This species has one generation per year and controls applied two to three weeks after peak adult flight (mid- to late-July), are effective in preventing the later larval damage.

Sod webworms overwinter as partially mature larvae that resume feeding in late April and early May. Lawncare providers often apply controls targeted at the summer generation larvae in July.

CURATIVE PROGRAMS
FOR NORTHERN TURF

Again, in our view, **an insecticide or other form of insect control should be used only when its use is justified**. The <u>focus of the **Curative Approach** requires</u> the turf user, service provider, owner or manager to <u>observe</u>, <u>survey</u>, <u>monitor</u>, <u>map turfgrass areas to locate, identify and assess</u> pest populations and/or symptoms of injury. Only when further population development and/or unacceptable damage is anticipated, should

control measures be employed. **The objective** is to prevent (avoid), stop, or at least limit (minimize) damage. **Always follow label directions**.

A thorough and complete knowledge of the seasonal spectrum of pests occurring at any location is essential for any approach to control.

In most insecticide tests, green June beetle grubs are more effectively controlled with curative treatments applied to the grass blades and upper thatch where these grubs surface to feed at night.

Infestations of black turfgrass ataenius often are not detected until the turf begins to show signs of water stress. A curative approach is needed <u>quickly</u> to save turf.

Greenbug aphid outbreaks are very difficult to predict. Controls applied when the aphids are first detected are effective.

CURATIVE PROGRAMS FOR NORTHERN GOLF COURSES

Primary Target - GRUBS

Knowing the complete spectrum of grub species common to a golf course is basic to the Curative Approach for preventing grub damage. <u>The seasonal occurrence of each of the life stages (eggs, larvae, pupae, adults) for each species is also necessary in order to know when to monitor for adults, and expect early symptoms of damage and/or larvae.</u> The times when these life stages are present may vary by three or more weeks between the northern and southern latitudes of a state. Another point to remember is that, **over time, the species spectrum may change**, therefore monitoring and correct identification of adults and larvae is important.

Regular Observation. The key to using the CURATIVE APPROACH successfully is to **prevent damage from grubs by early detection** (before significant damage occurs). This is accomplished by regular (daily) observation for early symptoms of grub infestation. **The Pest Spectrum Calendar for the golf course should indicate when to begin looking**. Symptoms such as areas of slight chlorosis (yellowing), unthrifty - droughty-

appearing turf, or wilt are too often **presumed** to be caused by localized dry spots or disease. This may be the case, however, if the life cycle of the grub species spectrum for the course (= <u>Pest Spectrum Calendar</u>) indicates this is also the time for early stages of a grub pest, <u>examination of the thatch and soil under such areas</u> may (or may not) reveal the early stages of a grub infestation.

The experienced eye of a golf course superintendent, who scouts the course daily, readily spots such symptoms. The key is, **DO NOT ASSUME** the cause of a symptom - **LOOK <u>UNDER</u> THE TURF**. The Diagnosis Section describing each of the major grub species in Chapter 3 of this book describes some of the symptoms.

Sampling and Mapping Grubs. *Grub infestations do not occur uniformly over a golf course*. Surveying and mapping the course to identify areas with and without potentially damaging populations is one way of locating areas that need to be treated instead of treating the entire course. <u>The methods and equipment useful in conducting such surveys are included in Chapter 9 of this book.</u>

When using the curative approach, early instar grubs are much easier to control than late instars. Early detection is essential!

Looking under the turf at the time that white grubs <u>might</u> be present is the only sure way to determine if they <u>are</u> present. First instar larvae are difficult to detect. A knife can be used for such inspections.

Among the **additional benefits** of surveying for grubs are: (1) identification of areas (fairways) prone to frequent infestations; (2) monitoring changes in the grub species by identifying all grubs as they are found; and, (3) reducing the total area requiring treatment and thereby the cost of treatment. The cost-effectiveness, in terms of time and money, as well as cost-benefit of such surveys, depends upon the perspectives of the golf course superintendent, financial considerations and course standards.

Monitoring Adults. Adult activity has never been successfully used to predict **WHERE** damaging grub infestations occur. However, knowing **WHEN** the different species of grub adults have flown can be used to predict approximately WHEN grubs can be sampled for or mapped. **Most annual grubs reach a size that is large enough to sample for within 50 days after the peak adult flight.**

Thresholds. Various publications on grub control often mention "population thresholds" or "action-non-action thresholds." Thresholds are the theoretical number of pest insects usually present in a given area (usually per square foot) before significant damage is experienced. The thought behind using thresholds is to reinforce idea that **application of a control material is not necessarily warranted simply because pests are present**. Pests must be present at populations high enough to eventually cause damage before application of a control.

Observing Japanese beetle adult activity is relatively easy. Grubs usually appear within 50 days of peak adult activity.

The **generally accepted level** for most grubs is 6 to 10 grubs per square foot before control is warranted. However, well maintained turfgrass with regular irrigation and fertilization can "tolerate" much higher grub populations. On the other hand, moles, skunks and racoons often find less than six grubs per square foot sufficient to dig up the turf in search of them.

Thresholds must be adjusted for each turf situation. For golf courses, damage in roughs is more tolerable than damage in fairways, and damage in fairways is more tolerable than damage on tees and greens. Likewise, some courses demand high turf standards while others may tolerate some periodic, localized damage.

Japanese beetle grub population at this site was 25 per square foot. The grub population in immediately adjacent turf (upper left) was similar but no damage was apparent.

The important thing in monitoring is to RECORD IT. Over time, the information recorded will be referred to many times when insect control and other turf management decisions are being made.

Keeping Records. Golf course owners and/or superintendents may, for various reasons, decide not to formally sample and map their golf course for grubs. Instead, they simply observe and spot-check the course regularly to determine if, when and where treatment is warranted. This approach can have substantial added value if the observations are recorded. A **bound notebook** dedicated to this purpose is best, though some computer programs allow for mapping and accurate record keeping.

Recordings are most useful when specific as possible, including information such as: (1) date and exact location of observation; (2) species identification; (3) sighting of first adults; (4) size (tiny, small, medium, large) of larvae; (5) symptom observed; (6) surrounding plant development, which is called plant phenology (bloom of trees, shrubs, weeds, etc.); and anything else that might be relevant.

Curative Insecticide Options. In late April and May, annual white grubs return to the upper soil level to briefly feed. Though grub damage may be evident at this time, skunk and racoon damage is usually the major problem. If necessary, trichlorfon is usually effective as a curative rescue treatment at this time, but the level of control may not be as high as with applications against early stage grubs.

As with most soil-inhabiting insects, insecticides are most effective when the small stages are the target. For most of the annual white grub species, first instar grubs are present near the soil surface from late July through mid-August. Applying an insecticide when larvae are present, followed with sufficient irrigation to move the insecticide to the grubs usually yields satisfactory results. A range of organophosphate, carbamate, pyrethroid and IGR insecticides are currently labeled for this purpose.

(left to right) Japanese beetle egg, 1st, 2nd, 3rd, instar grub, pupa and adult. The first and second instar larvae are most susceptible to curative controls.

First instar grubs molt into second instars which feed from late August through September. These middle stage larvae are also quite susceptible to control, but turf damage may begin to appear in mid-September as larvae mature. From late September to October, most of the annual white grub species have reached the third instar stage and are 40 to 60 times the body weight of the newly hatched, first instars. By this time significant damage may be evident and skunks, racoons and armadillos often dig up infested turf to feed on the grubs. Insecticide applications made at this time often yield poor control. Insecticides known to have rapid action and are least affected by thatch binding may be effective.

Biological Control Options. Commercial preparations of the insect parasitic nematode, *Steinernema carpocapsae*, have been marginally effective against Japanese beetle, masked chafer, Oriental and Asiatic garden beetle grubs as a curative treatment. This nematode and *S. feltiae* have not performed well or consistently against European chafer grubs. *Heterorhabditis* spp. and *S. glaseri* nematodes have been more effective in controlling various white grub species.

Nematodes should be applied when grubs are in their second instar. A minimum of 1/4-inch of irrigation before and after application increases effectiveness.

Primary Target -
BLACK TURFGRASS ATAENIUS

The presence of BTA or *Aphodius* **adults** on greens or in mower baskets **does not necessarily mean a damaging population of larvae will occur.** There is no known way of predicting the occurrence of larval infestations.

Larvae are the target of a curative approach to control of BTA and *Aphodius*, and the **objective is to kill them before the turf is damaged.** Fairways, tees and all putting surfaces should be examined daily from June through July for **evidence of wilt and/or yellowing** that precedes damage. Wilt will occur despite irrigation. A golf course cup cutter is an effective tool for examining areas showing symptoms. Grubs are usually found at the thatch-soil interface; however, smaller, early-stage larvae may be in the upper thatch.

Early symptoms of BTA or *Aphodius* infestations **are often misdiagnosed** as disease or localized dry spots. Only examination of the thatch and soil under turf showing such symptoms will identify the cause. **Misdiagnosis or late diagnosis can be costly.**

If turf wilts in June or July, immediately check to see if it is simply drought or BTA grubs. A cup cutter is a handy sampling tool.

Thresholds. Various thresholds of larval populations have been proposed for BTA and *Aphodius*; the most common is 50 per square foot (=5 per cup cutter sample). Actually, the level of fertility, irrigation, incidence of disease and overall health of the turf determines the level of infestation that can be tolerated. **"Healthy" turf can tolerate much higher populations than 50 per square foot.**

Insecticides. When observation determines that treatment is needed, biological and insecticide treatments are options. If the insecticide option is chosen, products such as trichlorfon, known to have rapid action and are least affected by binding to thatch, are known to be effective. **Irrigation immediately after application (before spray applications dry)** is important to achieving maximum control.

Natural Control. Examination of turf one to two weeks after application of insecticide sometime reveals considerable numbers of live but milky grubs. Apparently, the insecticide kills healthy non-infected grubs because a consequence of infection by the milky disease bacterium is that the grubs stop feeding and cannot ingest the pesticide. The presence of these living, but diseased grubs **may lead to the false conclusion** that the insecticide treatment was ineffective. Surviving larvae

Milky disease infected BTA grubs appear very white without the dark areas of the gut showing.

should be examined carefully for milky disease. **The more infected grubs seen, the better** since the dying grubs deposit millions of bacterial spores that will suppress future infestations.

Biological Control. Two insect parasitic nematodes, *Steinernema glaseri* and *Heterorhabditis bacteriophora* are labeled for curative control of BTA larvae. Control with *H. bacteriophora* has been good but variable with *S. glaseri*. Though these nematodes are not labeled for control of *Aphodius*, *H. bacteriophora* may provide control.

Primary Target - CUTWORMS, ARMYWORMS & SOD WEBWORMS

Often the **first damage** of the year (April) to golf greens may be **bird probes** caused by starlings feeding on overwintered sod webworm larvae and black turfgrass ataenius adults. The fecal droppings left by these birds commonly contain evidence of these two insects. When necessary, treatment with an insecticide (i.e., chlorpyrifos or a pyrethroid) to kill the larvae, stops further damage. An added benefit of such an application is that the previously described "sand cap covers" constructed by the overwintered sod webworm larvae (Chapter 8, page 98) **do not occur.**

Monitoring. A regular program of monitoring tees, greens and fairway areas for evidence of cutworm, armyworm and sod webworm infestations and applying treatment, only when larvae are

Trails in the morning dew indicate that some insect has moved across the surface. Soap flushing will determine if it was cutworms.

Treating the area around greens and tees can assist in reducing reinvasion by cutworms, armyworms, sod webworms and also provide control of other "secondary targets" such as billbugs.

present and/or damage seems eminent, is an essential part of a **Curative Approach**. Monitoring includes daily observation of the turf to look for larvae, damage and/or evidence of birds (especially starlings) probing for larvae. Probing damage to greens is usually intolerable, however, some golf course superintendents consider it a form of **natural / biological control on fairways**. Early morning observations of trails left in the dew where larvae have moved from one location to another is also an indication that closer inspection is needed.

An effective method to determine if larvae are actually present is to **apply a solution of liquid detergent** and water (two tablespoons of Joy® dishwashing detergent in two gallons of water) over a one-square-yard area to flush larvae to the surface. Sampling should be done every 14 to 21 days to detect early stage larvae, before significant damage is evident. This is also a good way to evaluate treatment effectiveness 3 to 4 days after an application. In our experience, this soap (Joy®) solution has not damaged turf though occurrence of some damage (yellowing) has been reported.

Applying a solution of liquid detergent is an excellent method to determine if cutworms, armyworms or sod webworms are present.

Large cutworm and armyworm larvae emerge quickly. Small sod webworms (above) emerge slowly, often after 25 minutes, and are difficult to see.

Control With Insecticides. Insect growth regulators (IGRs), spinosyn, and a broad range of insecticides effectively control cutworms, sod webworms and armyworms when a monitoring-based, **CURATIVE APPROACH** is used. **The principle** of control is to **apply the treatment and omit irrigation** for as long as practical (and permitted by the label) to allow the target insect to consume the treated foliage and contact the treated surface (= **The Target Principle**). Keep in mind, cutworms and sod webworms feed mainly at night and armyworms may feed day and night. Applications should be made as late in the afternoon as possible.

Curative programs are most successful when directed at early stage larvae. When treating greens and/or tees, **we strongly recommend** that the application extend one sprayer boom or spreader width beyond the affected area to control larvae that later may move onto the surface. This practice also helps control important "Secondary Targets" such as billbugs.

Biologically-Based Curative Controls

Insect Parasitic Nematodes. Products containing the insect parasitic nematodes, *Steinernema carpocapsae*, *S. feltiae*, or *Heterorhabditis bacteriophora* are available for control of these turf damaging caterpillars. Curative applications of these nematodes have given variable results and may not provide control (cessation of feeding) as rapidly as needed.

BT. Products containing the toxin derived from the bacterium, *Bacillus thuringiensis* (BT), are commercially available and adequately control caterpillars, especially armyworms, if enough material is ingested by young larvae. Use of these products has had mixed results in turfgrass management because of treatments being applied to a population consisting of large and medium sized larvae. Use of this microbial toxin is most effective when the larvae are in the first to third instar. Unless required by the label, irrigation should be omitted for at least 12 to 24 hours after application. Larvae affected by BT toxins rarely are found on the turf surface.

Azadirachtin. Azadirachtin is the active ingredient in a commercially available insect growth regulator (IGR) derived from extracts of the neem tree. Contact and/or ingestion **causes the insect to molt abnormally**. Both larvae and pupae are affected. This IGR has been used successfully for curative control of cutworms, armyworms and sod webworms on golf turf. Unlike with standard insecticides, **affected caterpillars do not come to the surface - they die in their burrows**.

Spinosyn. Spinosyn A & D is the active ingredient in a **new pesticide class** derived from a naturally occurring soil microbe and has been effective in controlling cutworm, armyworm and sod webworms.

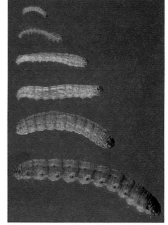

Black cutowrms have five to six larval instars. The first three are the best targets for curative control by biologically-based materials.

Treatment is most effective when applied before significant damage occurs (when larvae are small). High rates of spinosyn should be used to kill larger larvae according to label instructions.

Spectrum of Northern Golf Pest Insects Controlled by the Curative Approach

While the focus of a curative approach to controlling cutworm or armyworm **is** these pests (= **Primary Target**), examination of the Pest Spectrum Calendar for the golf course probably will show that <u>other pests occur at the same time on the site</u> (= **Secondary Targets**), and therefore may also be controlled or, at least, suppressed. Our principle is: **Insect pests of turfgrass rarely occur one at a time at any one time**.

For example: A curative application for control of cutworms in <u>June</u> could also provide control of <u>black turfgrass ataenius larvae</u> in the greens and <u>billbug larvae</u> in the area outside the green or tee if also treated. A cutworm treatment in <u>July</u> may also control <u>first generation sod webworm larvae</u>, kill <u>newly emerged black turfgrass ataenius adults</u>, and with careful selection of insecticide (halofenozide, trichlorfon), <u>early stages of certain grubs</u> (e.g., masked chafer).

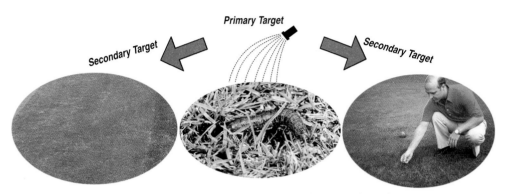

The target may be black cutworms, but black turfgrass ataenius grubs (early damage at left) and/or billbug larvae (damage at right) that are present at the same time may also be controlled or suppressed.

Notes

CURATIVE PROGRAMS FOR NORTHERN LAWNS

In our view, an insecticide or other form of control should be applied only when its use is justified. For a CURATIVE PROGRAM, the **justification should be the actual presence** of a potentially damaging population of pests. Assembly of a **complete Pest Spectrum Calendar** to know when to look, survey, or sample pest populations together with the color photographs, and diagnostic cues in this book will assist in turf analyses. Chapter 9 of this book provides useful information on methods and equipment for assessing pest population status.

For lawn service providers, training service delivery personnel to RECOGNIZE, RECORD and REPORT (and investigate if possible) symptoms of pest infestation on EACH PROPERTY serviced should help the business and client.

Primary Target - GRUBS

Insecticides. Occasionally, damaging populations of **overwintered grubs** are discovered in April or May. Such infestations mainly consist of third instar larvae that are difficult to control. **Treatment should be withheld until the population is in the upper two inches of soil**. Applications of trichlorfon or diazinon followed by thorough irrigation has provided adequate control.

Sampling for and identification of most grub species is possible beginning in early to mid-August depending on the grub species present. A broad range of effective insecticides are labeled for grub control at this time. The **primary reason for inadequate control is that the insecticide did not reach the activity zone of the grub**. Uniform distribution

Large grubs located below a thick thatch layer are difficult to reach with ANY control material.

and adequate post-treatment irrigation are essential to move the insecticide into the activity zone.

Applications made from late September to October yield poor control. If necessary, trichlorfon or diazinon may provide some control.

Biological Controls. Our experience with biological controls as a curative treatment for grubs has shown them to be variable to minimally effective.

Green June beetle grubs are relatively easy targets for curative controls because the grubs come to the surface to feed at night.

Primary Target - BILLBUG

The **target** of a curative program for control of billbug **is larvae** feeding in the stems and/or the crown. In northern regions, this generally occurs from mid- to late June but the time can vary considerably with location. **Timing of application is critical**. Therefore, if this time is not known, only careful observation and experience can determine when larval crown feeding occurs.

Once the larvae move into the soil to feed on roots and rhizomes, control is more difficult to achieve. Insecticides labeled for grubs and billbug larvae are the best option at this time. **NOTE:** Some products are labeled for billbug **adults** and may not be effective against **larvae**.

Biological Controls. Our experience with biological controls for control of billbug larvae have shown that the insect parasitic nematodes, *Steinernema carpocapsae*, *S. feltae* and *Heterorhabditis bacteriophora* are effective if label instructions (primarily pre- and post-treatment irrigation) are followed. Treatment is most effective when larvae are in the crown of the plant.

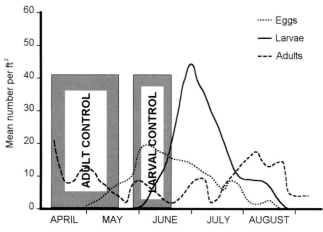

Over most of its range, curative control of bluegrass billbug larvae can be achieved from late May into the first two weeks of June. After that time, the larvae tend to move deeper into the soil-thatch zone and become a more difficult target to reach.

Curative control of bluegrass billbug can be achieved when the larvae are still in the stems or at the crowns.

Primary Target - CHINCH BUG

Infestations of chinch bug and billbug are especially damaging to turf during hot-dry periods of summer when turf may become dormant. In areas where two generations of chinch bug occur, both generations may require curative treatment but control of the first generation is often sufficient. ***Treatments should target nymphs before significant damage occurs***. A broad range of insecticides are labeled for chinch bug control.

Biological Controls. Biological controls, especially insect parasitic nematodes and *Beauveria* fungus applications, are <u>variable to minimally effective</u> as curative controls for chinch bug.

Cultural Control. Regular irrigation often suppresses chinch bug development by encouraging natural *Beauveria* fungus outbreaks. When large areas are damaged, consideration should be given to overseeding the entire lawn with endophytic ryegrass or turf-type tall fescue. Though technically a Preventive Approach, establishment of these grasses in the lawn will tend to suppress future chinch bug populations.

This lawn, on a south facing slope, is an excellent site for hairy chinch bugs. If this population is not treated soon, the turf will die and require renovation.

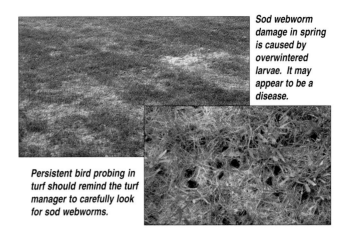

Sod webworm damage in spring is caused by overwintered larvae. It may appear to be a disease.

Persistent bird probing in turf should remind the turf manager to carefully look for sod webworms.

Primary Target - SOD WEBWORMS

Though several species of sod webworm may infest lawns throughout the growing season, first damage often occurs in spring (**April and May**) and is caused by overwintering larvae. Spring damage is often accompanied by probe marks left by starlings searching for the larvae. Summer and early fall generations also cause damage that can appear as general thinning of the turf. Starlings also persist in probing for larvae at these times.

If sod webworm activity is suspected, separate the turf stems with your hand to search for the presence of webbing and fecal pellets (frass). If the fecal pellets are greenish in color, the caterpillars are likely still active and curative controls will stop their activity.

Sod webworm larvae feed on grass leaves and stems (often under thatch), chewing them off just above the crown. <u>The principle of **CURATIVE CONTROL** of these insects is to treat the turf surface and allow time for the larvae to consume the treated foliage and contact the treated surface</u>. Therefore, unless required by the label, ***liquid applications of insecticide should not be irrigated in***. A broad range of insecticides including BT, azadirachtin, and maximum label rates of spinosyn are effective as curative controls for sod webworm.

CAUTION: Summer symptoms of sod webworm and earlier bluegrass billbug damage are often confused, especially when sod webworm adults are observed. ***Examine turf carefully to confirm diagnosis.***

One species, the ***cranberry girdler***, has a subterranean habit of feeding on turf roots and rarely feeds on green tissue. When significant populations of this species occur (***usually August to***

Sod webworm larvae are often difficult to find in turf, but their green fecal pellets (frass) are much easier to detect in the upper layers of thatch.

September), insecticides that provide curative control of grubs and are also labeled for sod webworm are most effective. Irrigation before and immediately after application helps obtain maximum effectiveness.

Biological Controls. The insect parasitic nematodes, *Steinernema carpocapsae* and *S. feltiae* can be effective, curative controls for sod webworm larvae **if** label directions are followed carefully.

Spectrum of Northern Lawn, Grounds and Athletic Field Pest Insects Controlled by the Curative Approach

Remember, while the focus of a curative approach may be one of the Primary Target Pests, examination of the Pest Spectrum and Target Calendar for a given area will probably reveal that other pests (= Secondary Targets) may be present at the time of treatment, and therefore, also controlled or suppressed, depending upon the Activity Spectrum of the Control Agent applied. The principle is: **Insect pests rarely occur one at a time at any one time.**

For example: A curative application made in mid-June for control of bluegrass billbug larvae could also provide control of summer generation sod webworm larvae as well as the first generation of chinch bugs.

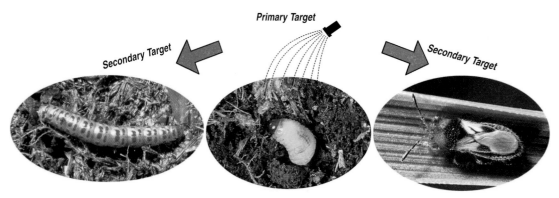

Primary Target

Secondary Target

Secondary Target

The target may be bluegrass billbug larvae, but overwintered sod webworm (left) and/or chinch bugs (right) that are present at the same time may also be controlled or suppressed.

Notes

PREVENTIVE PROGRAMS
FOR SOUTHERN TURF

Bermudagrass, zoysiagrass, centipedegrass, bahiagrass and St. Augustinegrass growing regions have **general pests** (e.g., mole crickets, grubs, armyworms, chinch bugs and fire ants) and **unique pests** (e.g., bermudagrass mite, spittlebugs, and ground pearls).

In our view, an insecticide or other form of insect control should be applied or implemented only when justified. The main reason to use a PREVENTIVE APPROACH is that past history of the turf being managed indicates a <u>chronic pest problem</u>.

By referring to your <u>Pest Spectrum and Target Calendar</u>, preventive treatments can be timed to simultaneously control the <u>Primary Target</u> and <u>Secondary Pests</u> as well, thereby reducing the number of applications necessary to achieve control.

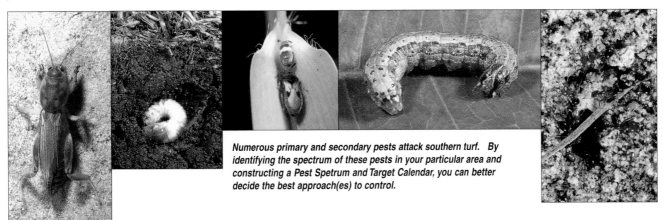

Numerous primary and secondary pests attack southern turf. By identifying the spectrum of these pests in your particular area and constructing a Pest Spetrum and Target Calendar, you can better decide the best approach(es) to control.

PREVENTIVE PROGRAMS FOR SOUTHERN GOLF COURSES

Primary Target - MOLE CRICKETS

The most difficult time to control mole crickets is late fall and early spring when adults are flying to relocate and mate. These adults may burrow deep in the soil profile - **below the Target Zone** - during cool or dry soil conditions, and therefore, are less prone to feed, which minimizes their exposure to control materials. Little can be done to PREVENT this movement and damage.

At sporadic times, usually associated with warm and rainy weather, adults move to the surface, tunnel extensively, fly in mass and mate. Research has shown that moist but not saturated sites with dense turf or weed growth are highly attractive to spring-active adults. <u>Such sites are where eggs will be concentrated</u>.

Mapping Mole Cricket Activity. In spring, areas where mole crickets are most actively tunneling, emerging, calling and digging back into the soil are where most of the eggs will be laid. A **visual inspection** of each <u>fairway</u>, <u>wetland margin</u>, and <u>managed roughs</u> should allow for easy detection of mole cricket "**hot spots**." With experience, the turf manager will learn to differentiate between light, moderate and extensive mole cricket activity.

Constructing maps of each fairway, drawn on standard paper and kept in a three-ring record book, <u>rough outlines of areas with extensive mole cricket tunneling can be drawn</u>. These identified sites will be areas at highest risk of having **significant turf loss** from mole cricket nymphal populations. **<u>Such sites are candidates for Preventive Control</u>**.

Subsurface Placement. In the recent past, insecticides with moderately-long residual activity (isofenphos and isazofos) were applied to high risk areas at the beginning of mole cricket egg hatch (usually late May to early June). With the discovery that **fipronil applied with subsurface placement equipment would provide season-long control of hatching mole cricket nymphs**, most other insecticides used as preventives have lost

This tunneling activity in April, following a rain, would indicate that this area is at considerable risk of being damaged from the new generation of nymphs. This site would be a prime candidate for Preventive Controls.

favor. Subsurface application of fipronil appears to have little activity against other soil-inhabiting insect pests of turf. Golf course managers using this tactic have experienced increased white grub damage as well as damage from animals digging for grubs. Apparently, these grubs were controlled or suppressed when the more traditional insecticides (e.g., isofenphos, isazofos) were used to control mole crickets!

photo: C. Anderson

Subsurface placement equipment is commonly used to place fipronil in the soil-thatch interface where it effectively prevents mole cricket nymph development. Such equipment is specialized and expensive.

Surface Application. When applied at mole cricket first egg hatch, surface application of **imidacloprid** adequately prevents mole cricket nymphal damage. The actual calendar date of this event varies considerably from south to north in areas where mole crickets occur.

For Example: Tawny mole cricket egg laying may begin as early as late March in south Florida, mid-April in north Florida and early May in south Georgia, and egg hatch occurs about 20 days later. *Generally, each major biological event in the life history of mole crickets is delayed by one week as one moves 100 miles from south to north.* Coastal and island areas can vary from this general rule.

Our opinion is that a single **surface application of imidacloprid** (at the highest label rate), made within the first three weeks of first egg hatch, should effectively prevent damage from tawny and southern mole crickets. Imidachloprid applied within this time period should suppress (if not control) the first new generation of cutworms, fall and true armyworms, and tropical sod webworm for 25 to 30 days after application, thereby eliminating the need for a surface insecticide treatment during this time.

A soap solution applied in areas where current mole cricket tunneling is observed is an effective method for capturing them, determining their age structure, or inspecting egg development of females.

Though neither labelled nor thoroughly tested at the time of this writing, **thiamethoxam** should also control mole cricket nymphs as well as a broad spectrum of secondary pests when applied at first egg hatch, as outlined for imidachloprid.

Other Surface Applications. Chlorpyrifos, acephate, or a registered pyrethroid may be applied to mapped areas that were determined to have considerable adult tunneling activity in April and early May. The insecticide is applied at egg hatch and every three weeks thereafter until egg hatching stops (usually after two to three applications). These applications also will control secondary targets such as cutworm, armyworm and sod webworms, **but will not effectively control grubs**.

Determining Onset of Mole Cricket Egg Hatch.
The key to successful use of imidacloprid (and possibly thiamethoxam) is to determine when mole crickets *in your area are ready to lay eggs*. This will require weekly sampling of adult mole crickets on the course, starting when spring flights and digging is prevalent. The soap irritant solution described in Chapter 9 will cause adult mole crickets to surface so that they can be carefully inspected. Capture 3 to 5 female mole crickets from several locations on the sites that were previously identified as "hot spots." with a sharp knife, open the abdomens of the female crickets and look developing eggs, a cluster of small, oval-shaped objects. If the eggs are flat to slightly oval and are soft and yellow-green in color, the female **is NOT** ready to lay eggs. **If the eggs are rounded, hard and dark yellow in color, egg laying will occur within 5 to 10 days**. First egg hatch normally occurs 20 days after egg laying. Again, a soap irritant solution can be used to detect the first instar nymphs.

Grub Control. April or May application of imidacloprid (and likely thiamethoxam) has enough residual efficacy to control **secondary pests** such as masked chafer or annual species of May/June beetle grubs that appear within 60 to 90 days after the application. Grub adults that lay eggs in August may not be controlled. Spring applications also appear to control hunting billbug.

Biological Controls. The insect parasitic nematodes, *Steinernema scapterisci* and *S. riobravae* have been touted as providing permanent, long-term preventive control of mole crickets. However, it is our experience that these nematodes, while often becoming permanently established in an area, do not produce the desired level of control expected by golf course managers, especially in high maintenance, irrigated turf. These nematodes may be useful in roughs, wetland sites and other lower maintenance turf areas.

photo: C. Anderson

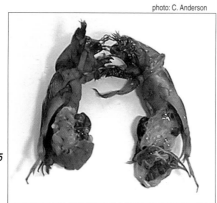

Female mole crickets must be dissected in order to observe egg development. Soft and yellow-green eggs (left) indicate that the female is not ready to lay eggs. When eggs are firm and dark yellow (right), egg laying will occur within 5 to 10 days.

Primary Target - **GRUBS**

If grubs are determined to be the **Primary Target**, and a **Preventive Program** is selected, the **first priority** is to determine which species or species complex is present (this information should be part of your Pest Spectrum and Target Calendar).

In many Gulf States, masked chafers and annual forms of May/June beetles are the most common grub pests. In Texas, Oklahoma and west, the southwestern masked chafer and annual May/June beetles are the common pests. The adults of these southern grubs usually fly and lay eggs when the rainy season begins or when summer rain fronts pass through. Flights of the southern and southwestern masked chafers are common in late July through August. The May/June beetles usually fly from May to August, depending on the species. Knowing which species is dominant and when it flies and lays eggs is **essential** for successful control timing.

May and early June applications of imidacloprid generally provide control of masked chafer and annual May/June beetle grubs except where the adults delay flight until mid- to late August (e.g., West Texas). This application should also control Secondary Targets such as mole crickets, cutworms, armyworms, tropical sod webworms and hunting billbugs.

Where **green June beetle** is also present, or **late flying masked chafers or annual May/June beetles** occur, **imidacloprid applications should be delayed until mid-July**. This treatment will provide season-long control of the grubs and suppress Secondary Pests such as cutworms, armyworms, tropical sod webworms and hunting billbugs. However, it is too late to provide mole cricket control.

May and early June applications of thiamethoxam should provide control of masked chafer and annual May/June beetle grubs and should also control Secondary Targets such as mole crickets, armyworms, cutworms, tropical sod webworms and hunting billbug. At the writing of this book, no data is available on the control of green June beetle with thiamethoxam.

June application of halofenozide has been shown to control masked chafer and annual May/June beetle grubs in July and August. Application at this time will also control Secondary Pests, such as cutworms, armyworms and sod webworms, that are present at the time of the application.

Primary Target - **FIRE ANTS**

Control of fire ants generally requires both **Curative and Preventive** approaches. Two very effective programs have been developed: the "**Two-Step**" and "**Ant-Elimination**" methods are satisfactory approaches for golf courses.

Two-Step Approach: This method requires an annual or twice-a-year application of a bait-formulated insecticide **first** applied over the entire turf area. The principle is to allow sufficient time for the fire ant workers to pick up these baits and take them back to the colony for distribution throughout the individuals. **Hydramethylnon baits provide control 3 to 5 weeks after broadcast**, while **fenoxycarb baits provide maximum mound control 4 to 9 months** after application.

One to three weeks after the bait is broadcast (to allow ants time to pick up the baits and take back to the colonies), the **second step** is to treat remaining, conspicuous or persistent mounds directly. Persistent mounds can be drenched, dusted, treated with granules, or aerosol injection with one of a range of insecticides registered for this purpose.

Once fire ants in an area have been brought under control, the Two-Step program can be used every year to prevent further buildup of new colonies. This is best done by applying the baits in the fall (September or October) and treating persisting mounds in the spring. If mounding becomes extensive, baits may be reapplied.

Ant-Elimination Approach: This approach is used where fire ants can not be tolerated and requires **broadcasting** a bait-formulated insecticide **and/or spreading granules around individual mounds**. After 2 to 3 days, a contact insecticide is applied to the entire area every 4 to 8 weeks to kill any foraging fire ant workers. When chlorpyrifos, acephate or a pyrethroid is used, Secondary Pests such as cutworms, armyworms and sod webworms occurring at that time will also be controlled. If applications are made when mole cricket eggs are hatching, many of the young mole cricket nymphs will also be killed.

The main principle of using fire ant baits is to let the ants have time to pick up the bait and transport it to the nest for further distribution throughout the colony. If transport of bait to the nest is disrupted by the application of other insecticides, the long-lasting effects normally obtained with baits will not be achieved.

Western (left) and southwestern masked chafers (right) often fly in July and August.

Green June beetle damage (right) can be prevented by applying controls when adults fly and lay eggs.

Individual mound treated with bait in the ANT ELIMINATION approach.

Fire ant baits contain slow acting toxins or insect growth regulators that disrupt reproduction or larval growth. The workers must be given time to collect the bait and take it into the nest.

Primary Target - OTHER PESTS

Cutworms, Armyworms, Sod Webworms and Tropical Sod Webworm. Cutworms, common armyworm, fall and yellowstriped armyworm, sod webworm and tropical sod webworm <u>rarely become the *Primary Target*</u> of concern on southern golf courses, therefore, **a preventive approach is not recommended**. We discourage applying an insecticide along with a fertilizer, herbicide, and/or disease control "as extra insurance" to control any larvae that <u>may be</u> present. Instead, <u>we favor a curative approach and application of a control (if necessary) when evidence of damage first appears or monitoring has indicated that caterpillar infestation risk is high.</u>

In many southern states, <u>fall and yellowstriped armyworm</u> outbreaks occur regularly in field crops and pastures. When this happens, state extension services often release "armyworm alerts." <u>We suggest using a daily visual monitoring program</u> whereby greens, tees and approaches are inspected for early signs of thinning or ragged leaf margins (grass blades often appear white). These are indicators of early armyworm and tropical sod webworm activity. Regular and **persistent bird feeding** (especially grackles, cattle egrets, ibis or starlings) in an area is also an indication that armyworms or tropical sod webworms are active. When necessary, it may be advisable to apply insecticide like chlorpyrifos or a pyrethroid every three to four weeks to prevent armyworm damage.

Applications of insecticides for control of armyworms also control other insects present (=Secondary Targets) such as fire ants, billbug adults, and young mole cricket nymphs.

Fall armyworm damage first appears as "frosting" and general thinning of the turf. When first noticed, a curative control product should be used.

Hunting and Phoenician Billbug. *Damage from these two billbugs is commonly misdiagnosed on bermudagrass* because it resembles damage caused by the disease, spring dead spot and delayed spring greenup. When careful inspection of the turf indicates signs of billbug activity (i.e., <u>chewed stolons</u>) or larvae are found, **a curative program should be used**. <u>If damage is extensive, a preventive approach should be considered for the next season for the affected area.</u>

Preventive application of imidacloprid, thiamethoxam, or halofenozide in May or early June will normally provide sufficient residual effect to kill billbug larvae that begin feeding in June through August. This approach reduces the population so few larvae will remain to overwinter and cause damage the following spring. Application at this time will also control or suppress mole crickets, grubs and larvae of cutworms, armyworms and sod webworms.

Regularly occurring hunting billbug damage can be prevented by applying imidacloprid, thiamethoxam, or halofenozide in May or early June.

Notes

PREVENTIVE PROGRAMS FOR SOUTHERN LAWNS

Primary Target - MOLE CRICKETS

When previous experience indicates mole crickets to be a perennial pest problem in a lawn, a **Preventive Control** program is warranted. At the writing of this book, fipronil is not registered for mole cricket control on home lawns.

At present, imidachloprid can be used in lawns as a preventive treatment. For most mole cricket zones, **imidacloprid should be applied within the first three weeks of mole cricket first egg hatch**. The calendar time of this period varies considerably from south to north in areas where mole crickets occur.

For Example: Tawny mole cricket egg laying may begin as early as late March in south Florida, mid-April in northern Florida and mid-May in South Carolina and egg hatch occurs about 20 days later. Generally, **each major biological event in the life history of mole crickets is delayed by one week as one moves 100 miles from south to north**. Mole cricket populations in costal or island areas are often ahead in development when compared with inland populations.

In our opinion, a **single surface application of imidacloprid** (at the highest label rate), made within three weeks of first egg hatch (as determined by occurrence of the first nymphs), should effectively prevent damage from tawny and southern mole crickets. Imidacloprid applied during this period should also control **Secondary Pests** such as masked chafer or annual species of May/June beetle grubs that can appear 60 to 90 days after the application. These spring applications also appear to control hunting billbug and suppress southern chinch bug populations.

Though neither labelled nor thoroughly tested, thiamethoxam should also control mole cricket nymphs as well as a broad spectrum of secondary pests when applied at the same time as outlined for imidachloprid.

Determining Onset of Mole Cricket Egg Hatch.

The key to successful use of imidacloprid (and possibly thiamethoxam) is to determine when mole crickets in an area are ready to lay eggs. State Cooperative Extension specialists often sample mole cricket populations and post notices when mole crickets are laying eggs. If this service is not available, **use a soap irritant solution** (described in Chapter 9) **to sample mole cricket adults, starting two weeks before the "normal" egg laying period for your area**. Capture three to five female mole crickets from different neighborhoods. With a sharp knife, open the abdomens of the females and look for a cluster of small, oval-shaped objects - developing eggs. If the eggs are flat to slightly oval and are soft and yellow-green in color, the female is NOT ready to lay eggs. **If the eggs are rounded, hard and dark yellow in color, egg laying will occur within 5 to 10 days** (see page 110 for photo). First egg hatch normally occurs 20 days after egg laying.

Lawn and landscape care providers who have used a Preventive Program for mole crickets should still **warn clients** that such applications will neither stop migration of nymphs from surrounding lawns nor stop incoming flights of new adults in late fall and early spring. If fall or spring migration of adults produces unacceptable tunneling, curative materials or baits may need to be applied.

Primary Target - GRUBS

If grubs (southern and southwestern masked chafer or annual May/June beetles) are the Primary Target, and a preventive approach is selected by home owners, lawn service client, or lawn service provider, **May to early June is the optimal time to apply imidacloprid**. In addition to providing seasonal control of grubs, the systemic activity of the material should also provide control of Secondary Targets such as hunting and Phoenician billbug larvae, and some suppression of early generation southern chinch bugs and fire ants. Control of first generation armyworm and tropical sod webworm larvae should also be achieved for 20 to 30 days after application, but will not prevent late season outbreaks.

Late June to early July application of imidacloprid provides season-long control of grub, from late flying green June beetles, *Phyllophaga crinita*, and will provide some control of billbug larvae. This late application will likely have little effect on the Secondary Targets, southern chinch bug and mole crickets, but development of armyworm and tropical sod webworm populations should be suppressed for 20 to 30 days.

May, June, or July application of thiamethoxam will control grubs in southern turf and secondary targets, similar to the activity spectrum of imidacloprid.

A June preventive application of halofenozide provides season-long control of masked chafers and *Phyllophaga crinita*. Control of *Phyllophaga latifrons* (the species common to south Florida) is limited. Infestations of the Secondary Targets, billbug, armyworm and tropical sod webworm larvae existing at the time of application may also be controlled with treatment at this time, but do not expect extended residual control.

Observed in September, the lawn on the left received no mole cricket treatment while the lawn on the right received a preventive mole cricket treatment in May.

Phyllophaga latifrons grubs commonly dig deep into the soil in September through November but armadillos will continue to dig for them. The only way to control these grubs is by using the preventive approach in May into July.

Application of imidacloprid, thiamethoxam, or halofenozide from July to early August will also control southern grubs as well as the green June beetle, but is too late for control of early generations of billbug, armyworm and sod webworms (***refer to your Pest Spectrum and Target Calendar***). The impact of treatment at this time on sod webworms that will overwinter is unknown

Primary Target - FIRE ANTS

Control of fire ants generally requires both Curative <u>and</u> Preventive approaches in Lawns. While the "***Two-Step***" or "***Ant-Elimination***" method can be used, <u>most home owners prefer the ant-elimination strategy</u> since it reduces the chance of having contact with fire ants in the landscape or home.

Two-Step Approach: With this method, an application of a bait-formulated insecticide is applied over the entire lawn once (spring) or twice a year (spring and early fall). <u>This method is more effective if entire neighborhoods cooperate with residents making similar applications</u>.

In **Step 1**, Hydramethylnon or fenoxycarb baits are broadcast over the entire lawn. Unfortunately, these baits act slowly on the fire ants within their nests. Hydramethylnon takes 3 to 5 weeks to achieve maximum effect, while fenoxycarb baits produce maximum control 4 to 9 months after application. After 1 to 3 weeks, many of the smaller fire ant mounds will have disappeared or become inactive. Where fire ant mounds are located in high use areas (nuisance mounds) or if larger mounds continue to persist, **Step 2** is to <u>treat nuisance and persistent mounds directly</u>. These mounds should be treated directly with drenches, dusts, granules, or an aerosol injection. A range of insecticides are registered for this purpose.

Once fire ants in an area have been brought under control, the Two-Step program can be used every year to prevent extensive build up of fire ant colonies. This is best done by applying the baits in the fall (September or October) and treating remaining mounds in the spring. If mounding becomes extensive, reapply baits.

Ant-Elimination Approach: In this approach, a bait-formulated insecticide is broadcast over the entire turf area or the bait is spread around individual mounds. After 2 to 3 days (to allow the foraging fire ants to pick up the bait), a contact insecticide is applies to the entire lawn. Depending on the number of foraging fire ants, the contact insecticide may need to be applied every 4 to 8 weeks. The main approach is to reduce mounds (baits) but keep encounters with fire ant workers minimized (contact sprays). When

Fire ant workers take a growth regulator-type bait into their nest. Inside, the bait will be ingested and eventually fed to the queens and/or larvae. Depending on the material, the queens may be sterilized or the larvae may not develop normally.

acephate or a pyrethroid is used as the contact insecticide, the Secondary Targets, cutworms, armyworms and sod webworms, will also be controlled. If these applications are made when mole cricket eggs are hatching, many young mole cricket nymphs can be killed.

Primary Target - OTHER PESTS

Billbug - ***Damage from billbugs, especially to bermudagrass or zoysia, is commonly misdiagnosed*** as disease, drought or winter damage. Hunting and Phoenician billbugs cause turf to die in patches and delay spring green up. <u>A Preventive Approach can eliminate such damage</u>.

Application of imidacloprid or thiamethoxam in **May or June** appears to kill both billbug adults and subsequent larvae. Application of halofenozide in June controls billbug larvae. Either of these applications should also **control the Secondary Targets**, white grubs and/or mole crickets. Application of carbaryl, diazinon or bendiocarb in **July or August** will also reduce billbug larval populations that can damage bermudagrass during fall and winter dormancy. White grubs present when these applications are made also will be controlled.

Hunting billbug damage is similar to disease spots or dog urine spots in a bermudagrass lawn. Billbug damage can be prevented by applying controls in May or June.

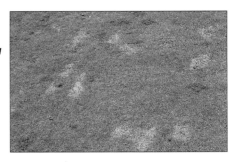

Southern Chinch Bug - At present, there are no registered insecticides that claim to provide long term control of southern chinch bug. However, our view is that **Preventive Programs** in which <u>imidacloprid or thiamethoxam are applied for control of Primary Targets</u> such as mole crickets or grubs should <u>suppress development of first generation southern chinch bugs</u>. Suppression of at least this generation would likely impact development of succeeding generations; how may remains to be seen!

Research in Florida has indicated that <u>entire lawns treated in May to July</u> with acephate or a pyrethroid <u>reduce southern chinch bug populations for most of the season</u>. Where smaller lawns are chronically reinfested from untreated surrounding lawns, a Preventive Program can be used whereby insecticide is applied every 3 to 4 weeks until cool weather arrives in late fall.

Fire ant mounds located in high use areas may have to be treated directly to eliminate the hazard. Such treatments often force the ants to relocate their mounds. The two-step and ant-elimination approaches provide better, long term control.

Insecticide Resistance. Populations of southern chinch bug resistant to certain insecticides, especially organophosphates and/or carbamates, have been identified in Florida and Texas and can likely occur where this pest is active all year. If an insecticide application does not appear to be effective, switch to another insecticide, preferably one in a different chemical category. In order to reduce the development of resistance, attempt to reduce the number and frequency of insecticide applications in a season.

An effective Preventive Control Program for southern chinch bug in St. Augustinegrass has been to use resistant cultivars, such as 'Floratam' and 'Floralawn' though 'Bitterblue', 'Sevelle,' and 'Floratine' have fair resistance. **Localized populations of southern chinch bug have overcome the resistance factors of 'Floratam.'**

Southern chinch bug damage can often be prevented by increasing irrigation frequency and improving coverage.

A material applied to control southern chinch bugs (left) or other pests, like armyworms or tropical sod webworms, in September through October is also likely to eliminate egg laying twolined spittlebug adults, thereby reducing their nymphal populations the following spring.

It has been our experience that where **irrigation** systems reach all parts of the lawn and irrigation is applied on a daily or every-other-day basis, natural *Beauveria* fungal infections appear to suppress chinch bugs below damaging levels (= Natural - Cultural Control).

Twolined Spittlebug - While common in bermudagrass and St. Augustinegrass, spittlebug is primarily a pest of centipedegrass. Where spittlebug is a chronic problem, a Preventive Program of applying acephate, carbaryl or a pyrethroid **at the time of spring greenup** has been effective in eliminating hatching nymphs.

Notes

Curative Programs
for Southern Turf

Again, our perspective is that **an insecticide or other form of insect control should be used only when its use is justified**. Because of the long southern turf growing season, the focus of the **Curative Approach** requires the turf user, owner or manager to constantly observe, survey, monitor, and map turfgrass areas to locate, identify and assess pest populations and/or symptoms of injury. When pests are detected and further population development and/or unacceptable damage is anticipated, employ control measures as may be necessary. **The objective** is to prevent (avoid), stop, or at least limit (minimize) damage.

A thorough and complete knowledge of the seasonal spectrum of pests, especially during times of abnormally wet, dry or hot weather, is essential for any approach to control.

CURATIVE PROGRAMS FOR SOUTHERN GOLF COURSES

Primary Target - MOLE CRICKETS

Knowing when (time of year) eggs, newly emerged nymphs, nearly mature nymphs, fall adults, or spring migrating and egg laying adult mole crickets occur is **essential** to success when using the **Curative Approach**. Your Pest Spectrum and Target Calendar must include these main life cycle phases. We strongly recommend periodic sampling (use a soap irritant flush to determine the stages present) and visual observation to determine the areas at risk of being damaged.

The **Curative Approach** involves broadcast applications of sprays or granules or bait formulations of insecticides. Broadcast Application of contact/stomach pesticides such as acephate, chlorpyrifos or pyrethroids is made when mole cricket eggs are hatching and again when nymphs are about half grown (no more than one inch long). To achieve maximum effect, the turf to be treated should be irrigated for several days before application to ensure that the soil is moist and the mole cricket nymphs are at the surface. When rainfall has been sufficient to move nymphs near the surface, pre-irrigation is not needed. To determine whether nymphs are near the surface, apply a light sprinkling (water can) of soapy water. If present, the small nymphs will "pop" out immediately. If it takes several minutes for them to surface, additional irrigation is needed to bring them closer to the surface before application.

The insecticides applied for curative control of mole cricket nymphs usually have little effect on white grubs, **but** other surface insects such as cutworms, armyworms, tropical sod webworms, and fire ants are often controlled or suppressed (= **Multiple Target Principle**).

After application of spray or granular formulations, irrigate lightly to move the insecticide to the TARGET nymphs. Inspect treated areas AT DAWN the following morning to determine treatment effectiveness. Affected nymphs will surface after being exposed to the insecticide. If few, or no, dead nymphs are found on the surface after the check for effectiveness, retreat with a different insecticide.

Bait-formulated insecticides are usually more effective once the nymphs are longer than one inch. Areas remaining moderate to highly infested after early applications to control small nymphs are candidates for treatment with insecticide-baits. Again, **irrigate PRIOR** to the application to move the nymphs to the surface, but **DO NOT IRRIGATE after applying bait**. Bait formulations are most attractive when fresh and not subjected to rainfall or irrigation. **Avoid** applying the bait if rainfall is expected the night after application.

Stopping turf damage from fall or spring migrating adults is difficult to achieve. The turf manager can do little to PREVENT this migration or damage. Baits can be used when control of adults is deemed necessary.

In our opinion (though not investigated), baits containing insecticides may also suppress turf-infesting larvae of armyworms and tropical sod webworms, as well as fire ants.

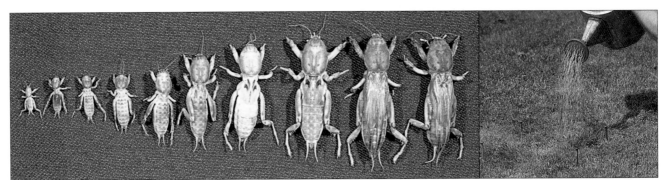

Eight nymphal instars of the tawny mole cricket, adult female and adult male. As their size increase, control is increasingly difficult to achieve. Curative controls work best against the first two to three instars. Baits are generally needed to control the large nymphs and adults.

A soap solution drench is an excellent method for determining size and numbers of mole crickets.

Primary Target - GRUBS

Where white grub damage has been an annual occurrence, a Preventive Approach is generally more effective than the Curative Approach. Southern grub species often feed deep in the soil profile, especially during dry periods, and therefore insecticides applied often do not reach the activity zone of the grubs, resulting in poor or no control.

If the **Curative Approach** is chosen and a significant grub population is discovered, the first step is to determine which species is present, their stage of development, and where the Target Grubs are located in the turf profile. If grubs are two to three inches deep in the soil, the turf should be irrigated for a few days to bring the "TARGETS" nearer the surface. The area should be resampled to determine if the grubs have actually moved within an inch of the soil surface. If the soil is already moist and the grubs are deep in the soil, irrigation will likely not move them closer to the soil surface. **DO NOT TREAT until the grubs are one to two inches of the soil surface**.

At present, trichlorfon is the main curative grub insecticide, but ethoprop, and carbaryl have also been effective. Irrigation immediately after application with as much water as can reasonably be applied (1/4- to 1/2-inch) helps move the pesticide to the TARGET. **However, never allow puddling!**

All insecticides labelled for control of turf insect pests can control a variety of other turfgrass insects (ethoprop is also a nematicide). If mole crickets, turf-infesting caterpillars, mites, scales, or billbugs are present, most will be controlled or suppressed by curative grub treatments (= Multiple Target Principle). **Carefully review your Pest Spectrum and Target Calendar** to determine what other targets are present and may be affected by a treatment.

NOTE: If skunk, raccoon, or armadillo digging is the actual concern, control of these animals by legal trapping, removal of nesting sites and food, etc., is often more effective than trying to control the white grubs they feed on.

Many superintendents seem to have an uncanny ability to anticipate infestations of fall armyworms and tropical sod webworms. They have often developed an "eye" for detecting armyworm egg masses on marker flags and signs. These managers also notice when tropical sod webworm adults appear in roughs or landscaped areas surrounding fairways.

Primary Target -
CUTWORMS, ARMYWORMS, SOD WEBWORMS and TROPICAL SOD WEBWORMS

Cutworm, sod webworm, and/or common, fall or yellowstriped armyworm damage on greens can occur any time during the season. Armyworm and tropical sod webworm damage to other golf course areas is usually most common at the onset of the "rainy season," but can also occur any time during the warmer months. Regular and persistent bird probing often indicates infestations of these pests.

Monitoring. A regular, daily program of examining tees, greens and fairways for evidence of caterpillar infestations is the key to success when using the Curative Approach. If damage is not evident, but **birds persist probing** these areas, use the soap flush (two tablespoons of Joy® dishwashing detergent in two gallons of water, sprinkled over a one square yard area) to determine what insects are present. Remember that birds will also probe for mole cricket nymphs. It has been our experience that this solution has not caused damage to turf, but it is always wise to test the solution in an inconspicuous spot before using it more widely.

Mole crickets or white grubs (above) may be the Primary Target of an insecticide application on a southern golf course. This application may also control Secondary Targets like: hunting billbug adults (left), fall armyworm or sod webworm larvae (center), and even fire ant workers (right).

Control With Insecticides. A broad range of insecticides (including insect growth regulators and spinosad) effectively control cutworms, armyworms, sod webworms and tropical sod webworm when a <u>monitoring-based</u> **CURATIVE APPROACH** is used. **Apply the treatment and <u>omit irrigation</u>** for as long as practical to <u>allow the target insect to consume the treated foliage and contact the treated surface</u> (= **The Target Principle**). While cutworms feed mainly at night, armyworms and tropical sod webworms may feed day and night. **Applications should be made as late in the afternoon as possible to provide maximum residues when larvae begin their night feeding**.

Early stage larvae are the best Target in Curative Programs. **When treating greens and/or tees, we strongly suggest also treating the turf beyond these areas to control larvae that later may move onto the surfaces as well as other Secondary Targets**.

Biologically-Based Curative Controls

Insect Parasitic Nematodes. Products containing the insect parasitic nematode, *Steinernema carpocapsae*, *S. feltiae*, or *S. riobravae* are available for control of caterpillars damaging southern turf. <u>Based on our experience, curative applications of these nematodes produce variable results</u> and may not stop feeding as rapidly as desired. <u>These nematodes also affect billbugs, and S. riobravae is known to seek out mole crickets</u> (= **Multiple Target Principle**).

The billbug <u>adult</u> at right was found dead after an application of insect parasitic nematodes applied to control cutworms. Upon breaking the billbug open, adult nematodes were discovered to have infested it! (= Multiple Target Principle).

BT. Control products containing <u>*"Spodoptera"* active strains</u> (*Spodoptera* is the genus of the fall and yellowstriped armyworms) of *Bacillus thuringiensis* have provided <u>good curative control</u> of armyworms, sod webworms and tropical sod webworm when applied to control early-stage larvae. These products are best applied when infestations are first noted and repeat applications may be needed every 4 to 5 weeks during the warmer part of the season. **Affected <u>larvae rarely surface</u>**. Caterpillar active strains of BT have no affect on other insects.

Azadirachtin. Azadirachtin is the active ingredient in a commercially available insect growth regulator (IGR) derived from extracts of the neem tree. Contact and/or ingestion causes the insect to stop feeding and molt abnormally. Both larvae and pupae are affected. <u>This IGR has been used successfully</u> for curative control of cutworms, armyworms and sod webworms in golf turf. Unlike with standard insecticides, **affected caterpillars do not come to the surface - they die in their burrows**. Because of the short residual activity of azadirachtin, few other insects are affected.

Spinosyns. Spinosyn A & D are the active ingredients in a new pesticide class derived from a naturally occurring soil microbe and <u>has been effective</u> in controlling cutworms, armyworms and sod webworms. Treatment is most effective when applied before significant damage occurs (when larvae are small). High rates of spinosyn can be used to kill larger larvae. Spinosyns affect other turf insects, but because of the short residual activity (less than five days) and poor ability to penetrate through thatch, turf-infesting caterpillars are primary targets.

Primary Target - **FIRE ANTS**

The "Two-Step" and "Ant-Elimination" methods discussed on page 111 are the best approaches. These methods use both curative and preventive principles.

In the **Two-Step Method**, the **first step** is to apply a bait-formulated insecticide <u>over the entire turf area</u> followed one to three weeks later with the **second step** which is to <u>treat nuisance mounds (ones located in high traffic areas) directly</u>. The **Ant-Elimination Method** requires bait-treating nuisance mounds and then treating the whole turf area with a general contact insecticide every 4 to 8 months.

When fire ant mounds suddenly appear in undesirable places, baits may be immediately applied around the mound. If more rapid elimination is necessary, select one of the insecticides registered for drenching, dusting, granule spreading, or aerosol injection of fire ant mounds. Acephate dusting or chlorpyrifos injection systems have produced very rapid individual mound elimination. When larger areas are treated, <u>Secondary Pests</u> such as cutworms, armyworms, sod webworms, and tropical sod webworms will also be controlled. If applications are made when mole cricket eggs are hatching, many young mole cricket nymphs will also be killed.

When using bait-formulated insecticides for fire ant control, the mound itself does not need to be treated. Generally, apply the bait around the mound or over the turf in the general vicinity of the mound so that the workers can pick up the bait and take it into the nest.

Primary Target - OTHER PESTS

Billbug - Hunting and Phoenician billbugs are difficult to control in bermudagrass and St. Augustinegrass using a Curative Approach because damage often does not appear until warm spring weather stimulates turf growth. ***Damaged turf is frequently misdiagnosed since it often appears as dead spots or areas with delayed growth***. At this time the larvae have finished development. However, if turf damage is discovered and billbug larvae are active in the upper soil, pre-irrigate to moisten the soil and apply an insecticide specifically registered for billbug larvae. Be sure to consult you Pest Spectrum and Target Calendar, since turf-infesting caterpillars and young mole cricket nymphs can often be controlled by the same insecticides registered for billbug larvae (= Secondary Targets).

Bermudagrass Mite - Curative control of this mite is difficult to achieve once the turf shows typical stunting and tufting symptoms. Application of a registered insecticide twice, at a three-week interval, can assist recovery of the turf. The Preventive Control

Stimulating new growth by watering and fertilization help mask bermudagrass mite (left) and bermudagrass scale (right) damage.

Approach of increasing fertilization and irrigation to encourage rapid new growth is generally recommended.

Bermudagrass Scale - At present, no insecticides are registered for control of this scale. The Cultural Approach of increasing fertilization and irrigation generally helps the turf outgrow damage from this pest.

Notes

CURATIVE PROGRAMS FOR SOUTHERN LAWNS

Justification for a **CURATIVE CONTROL** program for insect pests in warm-season lawns should be actual presence of a potentially damaging pest population. A **Complete Pest Spectrum and Target Calendar** for the area along with the color photographs in this book will assist in determining when such populations may occur.

The decision to TREAT or NOT TREAT depends upon:
(1) PERSPECTIVES of the person(s) making the decision(s);
(2) FINANCIAL considerations;
(3) TURF QUALITY STANDARDS desired; and,
(4) the present and past PEST SPECTRUM.

Primary Target - MOLE CRICKETS

The Curative Approach to mole cricket control in lawns is **not a good option** because, by the time it's discovered, the population usually consists of nearly mature nymphs that are difficult to control.

When small nymphs are detected, application of a registered **insecticide followed by irrigation is effective**. Depending on the insecticide selected, other pests (= Secondary Targets) such as white grubs, fire ants, turf-infesting caterpillars, billbugs, chinch bugs and spittlebugs may also be controlled or suppressed.

If medium to large nymphs are present, an insecticide formulated bait will also lower the population below damaging levels. **Irrigation should be suspended for two days following an insecticide-bait application**.

When large mole cricket nymphs and adults are tunneling in the fall or early spring, little can be done other than using an insecticide-bait in an attempt to lessen the activity. Such applications will neither stop migration of such large nymphs from surrounding lawns nor stop incoming flights of new adults.

Biological Controls. "Claims" have been made
that application of the insect parasitic nematodes, *Steinernema scapterisci* and *S. riobravae* for mole cricket control can "establish" these nematodes to provide both short-term and long-term control. Our opinion is that application of 0.5 billion infective juveniles per acre, immediately followed by 1/4 to 1/2 inches of irrigation occasionally provides curative control of medium to large mole cricket nymphs.

Fire ants can be beneficial since they prey on other insects such as this tawny mole cricket male. Apparently, application of imidachloprid or thiamethoxam disrupts normal mole cricket behavior making them more vunerable to attack by various predators and parasites.

Primary Target - GRUBS

The species or species complex must first be identified. Since mixed populations of masked chafers, annual May/June beetles and the green June beetle can be present, knowing which are present, what stage they are in (first, second or third instar) and where they are located in the soil profile can help determine the most appropriate Curative Treatment and application timing.

Occasionally, annual May/June beetles (i.e., *P. latifrons* in Florida) lay eggs in April into May that produce second instar larvae by late June into July while southern and southwestern masked chafers may lay eggs in July and into August that result in second instar larvae in September. This 40- to 60-day difference can require a curative grub insecticide application for May/June beetle larvae in July (which misses the masked chafers that have not yet hatched) and another in early September for the masked chafers. Fortunately, in many warm-season turf zones, the annual May/June beetle species and masked chafer adults lay eggs about the same time, so both grubs can be controlled by a single application. **Thorough and immediate post treatment irrigation is a MUST!**

When animal digging (skunk, raccoon, armadillo) is the main cause of concern (i.e., grub damage is not evident), an application of diazinon or trichlorfon is often sufficient to discourage digging. If this does not work, legal animal trapping may be the only solution. Be sure to check with local wildlife officials for regulations concerning trapping.

Most insecticides registered for curative grub control also affect a variety of other targets (= Secondary Targets). Consult your Pest Spectrum and Target Calendar as well as the pesticide label to determine which other targets may be controlled (= Multiple Target Principle).

Areas such as this street lawn are often not irrigated properly and chinch bug damage occurs.

Fire ant mounds that suddenly appear in a lawn are eliminated by applying a bait around the mound. Where people may contact the ants, rapid elimination may require a direct insecticide application.

Primary Target - CHINCH BUGS

Damage from southern chinch bug is similar to symptoms of drought stress or disease. Therefore, **NEVER ASSUME** that there is a sole cause of turf yellowing, browning or thinning. When damaged lawn turf is noted, ***carefully inspect the turf to identify the TRUE CAUSE***.

If damaging populations of chinch bug are noted, a broad range of registered insecticides are available. In shorter-cut grasses (i.e., bermudagrass or zoysiagrass), little, if any, post-application irrigation is needed. In higher-cut grasses (i.e., St. Augustinegrass), 1.5 inch of irrigation or more may be needed to move the insecticide into the thatch (= Target Zone). In general, liquid applications, applied in high volume (3 gal./1000 ft″ or more) have been more effective than low-volume or granular applications.

Most insecticides registered for chinch bug control also control many of the other leaf, stem and crown-inhabiting pests (= Multiple Targets). Mole cricket nymphs may also be controlled or suppressed depending on the time of application. Review your individual Pest Spectrum and Target Calendar for other pests that may be present and controlled by the chinch bug application.

Insecticide Resistance. Populations of southern chinch bug in Florida and Texas have been found to be resistant to certain insecticides, especially organophosphates and/or carbamates. If the application of an insecticide in one of these classes has not produced satisfactory control, one of the new pyrethroids may be effective.

Irrigation. Infestations of southern chinch bug are often related to drought conditions or nonuniform irrigation. Populations can often be brought under control by correcting nonuniform irrigation distribution and beginning regular irrigation (= Cultural Curative/ Preventive Approach). Population decline and turf recovery are usually the result of a combination of chinch bugs being infected with the fungus disease, *Beauveria*, and vigorous turf growth.

Primary Target - FIRE ANTS

While the combined Curative and Preventive Approach of the **Two-Step Method** is considered the most effective approach (bait application once or twice a year - prevention - then treat persistent mounds - curative), many residents prefer the **Ant Elimination Method** (apply a bait directly to mounds, wait 3 to 5 days, then treat the entire lawn with a contact insecticide every 4 to 8 weeks). Where a fire ant mound suddenly appears in an undesirable place or a mound persists, and rapid elimination is desired, select one of the insecticides registered for drenching, dusting, granule spreading, or aerosol injection of fire ant mound(s).

When entire lawns are treated with contact insecticides, other leaf, stem and crown pests are often controlled (= Secondary Targets). Control of leaf-infesting caterpillars, chinch bug and billbug adults, as well as nuisance fleas is especially evident when applications are made on a regular basis (every 4 to 8 weeks) (= Multiple Target Principle).

Primary Target - OTHER PESTS

Billbug - Hunting and Phoenician billbugs usually cause damage in bermudagrass, zoysiagrass or St. Augustinegrass when these grasses are growing slowly during drought or cool conditions. If billbug larvae are detected in damaged turf, an application of an insecticide labeled for billbug larvae, followed by sufficient water to move the insecticide into the Target Zone, will decrease further damage. At the same time, any agronomic practice (increasing fertility or irrigation) that increases turf growth will help the turf outgrow the billbug damage (= Cultural Control).

Twolined Spittlebug - Refer to your Pest Spectrum and Activity Calendar to determine when spittlebug eggs usually hatch for your region. Two to three months after this predicted hatch period, inspect the turf (mainly centipedegrass, but occasionally bermudagrass and zoysiagrass) by spreading the turf canopy in several places to determine if spittle masses are present at the bases of grass stems.

Application of an insecticide labeled for control of this insect followed by sufficient irrigation to move the material to the thatch surface usually reduces spittlebug nymphal populations to tolerable levels.

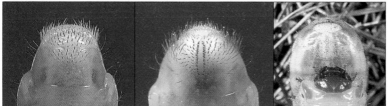

Do not try to identify grub species by their general size and form. The only sure way to identify them is to look at the raster pattern (see page 19 & Chapter 3). While there are some 50+ species of grubs in some warm season turfs, masked chafers (left), annual May/June beetles (center) and green June beetles are the most common grubs found in southern lawns.

121

Armyworms and Tropical Sod Web-

worm - In warm- season turf, fall and yellowstriped armyworm or tropical sod webworm infestations <u>can occur any time during the warmer months</u>, but they usually occur shortly after the rainy season begins. During these periods, insecticides labeled for their control may be necessary every 3 to 4 weeks. Increasing fertility and irrigation may result in the turf outgrowing the damage and decrease the need of insecticide application (= Cultural Control).

Multiple Target Principle - *Insecticides applied for curative control of billbugs, spittlebugs or turf caterpillars, usually control a variety of leaf, stem, crown and even soil-inhabiting pests. By reviewing your Pest Spectrum and Target Calendar, these Secondary Targets may also be controlled with a single, <u>well-timed application</u>.*

Notes

Integration of Control Approaches

Authors' Perspectives

Until recently, chemical and biological approaches to insect control were generally considered incompatible because many insecticides, fungicides and herbicides had, at least, some detrimental effect on natural controls such as: **Predators** (spiders, mites, ground beetles, big-eyed bugs, etc.); **Parasites** (wasps and flies); and **Pathogens** (bacteria, fungi, nematodes). Supplemental introductions of these natural agents (=Biological Controls) were also, to varying degrees, negatively impacted by some pesticides.

Recent research (from California, Kentucky and Ohio) on the impact of the chloronicotinyl insecticide, imidacloprid, and a bisachylhydrazine, halofenozide on natural or biological agents has OPENED THE DOOR TO INTEGRATING CHEMICAL, NATURAL, AND INTRODUCED BIOLOGICAL CONTROL.

Example 1.

Imidacloprid is systemic and primarily toxic to insects that ingest treated thatch or soil, plant parts or fluids. Insect predators and parasites do not feed on plant parts or thatch and soil, therefore, are not killed. Furthermore, recent research has confirmed that once ingested, the insecticide **modifies the behavior of the pest** to the extent that it loses its ability to defend against natural or introduced enemies. This effect is now being called "**synergistic.**"

Considering the 120-day half-life of imidacloprid in soil, this "**synergistic**" effect will affect any susceptible pest that occurs throughout the turf growing season. We think further research will show that imidacloprid is not only compatible with natural and introduced biological controls, but may result in increased populations of these organisms!

Example 2.

Halofenozide is an IGR (Insect Growth Regulator) that accelerates the molting process in insects and thereby causes death. Though somewhat systemic, ingestion of treated soil or thatch (with a half-life of 100+ days), and other plant parts is apparently the primary mode of action. Recent studies in Kentucky showed that halofenozide caused no reduction in the abundance of any beneficial invertebrates. Additional research in Indiana, indicated that application of this IGR does not inhibit or kill insect parasitic nematodes, but synergism has not been demonstrated.

Evaluation of Your Approach
and Control Program

Evaluate YOUR APPROACH (Preventive, Curative, Tolerance) and ***CONTROL PROGRAM(S)*** for destructive turf insects by carefully considering the program(s) YOU USE in relation to the ***Pest Spectrum and Target Calendar*** you developed (we hope!) earlier in this chapter. Consider:

(1) Is the COMPLETE SPECTRUM of Primary and Secondary Pests being considered, or is the focus mainly on **one** Primary Pest?

(2) ***Can one or more applications be OMITTED?***

Meaningful evaluation requires in-depth knowledge of the control agents and cultural methods employed. This information is too extensive for us to present in this book. However, product labels, technical information sheets, results from university trials, turf conferences, state extension specialists and the experience of nearby turf managers and/or service providers are always available.

The effectiveness of a control agent or method can range from one to many pest insects. Knowing which pests AND pest STAGES are or are not controlled should form the basis for selecting the products used and optimal time for treatment to achieve maximum benefit.

Attention to the MULTIPLE TARGET PRINCIPLE which includes consideration of PRIMARY and SECONDARY TARGETS can reduce the number of treatments necessary to achieve the turf quality standards desired.

Notes

Nothing beats simply using the "hands-and-knees" method for determining symptom cause!

Equipment, Methods and Approaches For Detecting, Surveying and Monitoring Turfgrass Insects

This Chapter

In this chapter, we present equipment, methods and approaches for detecting, surveying, monitoring, identifying, and diagnosing problems caused by destructive turf insects. When used before and after employment of control measures, the information in this chapter also can be useful in determining treatment effectiveness.

Perspectives

Observation, surveys, monitoring, mapping turf areas to locate, identify and assess pest populations and/or symptoms of injury are fundamental to a successful insect control program. Of these, **OBSERVATION** and **RECORDING** when and what is observed is most basic.

Observation requires development of a **keen sense of awareness, symptom sensitivity and curiosity** any time a turf area is visited. Memories fade, therefore observations should be recorded. Entry into a computer is OK, but, a bound notebook is best. Over time, these collective records and observations contribute important specific details to the **Turf Pest Spectrum and Target Calendar** of an area or site, and improves the observer's ability (together with this book) to know **when** to expect pests in the spectrum to occur.

Equipment

Detection and identification are the first and most important steps in dealing with turfgrass problems. Regular monitoring for pests provides information on **where** pests are located and **when** pests reach levels that need a control action. Several pieces of equipment are useful in detection and monitoring.

Golf Course Cup Cutter. The standard golf course cup cutter (4.25-inches in diameter) is a convenient tool to survey for grub and other soil inhabiting insect infestations in golf courses and home lawns. If care is exercised, sampling can be done with minimal damage to the turfgrass area. Once removed, samples can be examined on the spot, then the soil and turf can be placed back in the hole made by the sampler. In cases where it may be desirable to remove the samples for detailed examination elsewhere or extraction in a Berlese funnel, the hole may be filled with a plug of

The standard golf course cup cutter can be used to sample grubs in all turf areas - sports fields, lawns, and golf courses. Remove a plug, split it open from the bottom to locate and count grubs, then carefully replace it. Complete destruction of the sample is usually not necessary.

similar turf from a nearby area. When the soil under the turf is dry, adding water to the sample hole before replacing the plug helps turf survival.

Source: *Cup cutters are available from golf course equipment and supply companies listed in the appendix of this book.*

Berlese Funnel.

The Berlese funnel is a device in which heat from an incandescent light bulb is used to force living insects and mites from samples taken in the environment they inhabit. Primary components are a steep-edged funnel fitted with a 1/4-inch screen on the inside, and standard jar cover and 1/2-pint jar containing ethyl alcohol (20 to 50%) at the base.

Samples of turfgrass are taken with a standard golf course cup cutter and placed, grass side down, on the screen ledge inside the funnel. The heat source (a 25 watt bulb) is then positioned two inches above the top of the funnel and held in place for 24 hours.

Moist samples require 48 hours for extraction. Insects forced out of the sample fall into the funnel and attached jar of ethyl alcohol below. About one inch of alcohol in the jar is sufficient. Since the collected material is preserved in alcohol, it may be examined at any convenient time and/or may be sent to an expert for detailed analysis.

Turfgrass samples taken for extraction should have an inch or less of soil attached because excess soil interferes with heat penetration. In addition, the Berlese system is not effective in removing insects from the soil portion of the sample because they are unable to escape before the heat or desiccation kills them. The device is most effective for insects that inhabit the thatch and soil surface (billbug adults, chinch bugs, sod webworms, cutworms, mites, springtails, etc.), or those on the grass blades (aphids, mites).

A 10X hand lens or larger magnifying glass is adequate for general examination of material collected, but detailed analysis requires the use of a microscope. Conversion to number of insects per square foot is achieved by multiplying the number obtained from a sample 4.25-inches in diameter by 10.15. Generally, those persons examining samples extracted with a Berlese funnel for the first time are amazed at the variety of insects and mites present.

This cup cutter sample can be placed in a Berlese funnel to extract the insects and mites that inhabit the soil surface and thatch.

Berlese funnels set up to extract insects and mites from turf/soil samples.

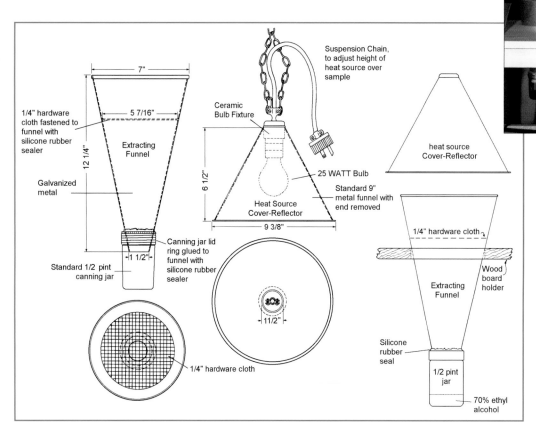

Plans for constructing a Berlese funnel.

Professional light trap used by entomologists to monitor night-flying insects, including turf pests.

Light Traps.

Professional light traps are basically black lights with baffles that deflect attracted insects into a container. Insects attracted to the light hit the baffles, fall in a funnel and into a container below. A plastic "pest strip" or similar device that releases a volatile insecticide is placed in the container to kill the insects soon after they enter. A cotton or cloth filled can containing a killing agent (ethyl acetate is preferred) can also be used. The layman or amateur should be aware that thousands of insects can be collected in just one night during the summer. Professional expertise is usually needed to analyze the material collected.

A **Simple Light Trap** for monitoring the flight activity of night-flying beetles such as May-June beetles, masked chafers, and black turfgrass ataenius is easily assembled. A rain-shielded 60 watt light bulb is hung three feet over a small child's plastic swimming pool, wash tub or similar container nearly full of water. Beetles are attracted to the light and fall into the water. Placement next to a light colored building helps.

Beetles on the surface are counted, recorded and removed each day. Regular tending of this simple light trap will indicate when first, peak, diminished and last flight activity occurs.

Source: *Professional light traps can be purchased from entomological supply companies listed in the appendices of this book.*

A typical mid-June light trap collection from one night shows hundreds of different insects, not all of which are turf pests.

Pheromone Traps.

Pheromones are the chemicals that insects use to find each other (sex or aggregation pheromones), plants use to attract pollinators (e.g., "floral" lure of Japanese beetle traps), or are otherwise involved with "communication." These pheromones can be manufactured and used in traps to monitor the activity of insect pests.

The Japanese beetle trap is the most commonly, commercially available pheromone trap. This trap is sold for control of Japanese beetle, but research has shown that the traps capture only 50 to 60% of the adult beetles attracted and more beetles are drawn into the area than would normally be present. If used, the trap should be placed well away from favored adult food host plants and used only to detect first, peak and ending flights of Japanese beetles. Record, remove and dispose of the beetles trapped each day.

Commercially available Japanese beetle "bag" trap.

Pheromones have been identified for Oriental beetle and green June beetle and May/June beetles, but these are not commercially available. Oriental beetle pheromone traps are placed on the ground since the beetles tend to fly within a foot of the turf surface.

Pheromone traps baited with sex pheromones for black cutworm, and fall and yellowstriped armyworms are commercially available and some golf superintendents use them to determine adult flights of these pests. We do not recommend general usage of these traps since there has been little research to show a relationship between adult flight and larval damage. Pheromone traps for cutworm and armyworms should be replaced or cleaned daily since trapped males appear to reduce the attractiveness of the trap to other males.

"Delta" trap used for cutworm monitoring (above) and male cutworms caught in the "sticky" trap bottom (right).

Source: *Some pheromone traps such, as the Japanese beetle trap, are available from garden stores. More specialized traps can be purchased from companies specializing in entomological or integrated pest management equipment listed in the appendices.*

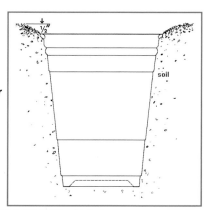

Diagram of a pitfall trap that can be used to monitor surface-inhabiting insects and mites.

Mole Cricket Sound Trap. The *male "chirp" of the tawny or southern mole cricket* can be reproduced electronically or through audio recording. When the "song" of either species is played at dusk and continued for two hours after dark, both females and males of that species are attracted. The speaker may be hung over a child's plastic wading pool, half filled with water, so attracted adults fall into the pool. Floating adults can be counted, recorded, removed and destroyed the next day. Made in the spring, such counts are useful in determining when migratory flights begin and stop. The best time map areas where mole crickets are tunneling and likely to lay eggs is when trap captures greatly decrease.

photo: B. Joyner

Two mole cricket traps, one with the tawny mole cricket and the other with the southern mole cricket "chirp" devices (the boxes suspended over the wading pools).

Source: *The electronic chirp devices were once available from the University of Florida, but we are not aware of a current source. Tape recordings are sometimes available from state cooperative extension (county agent) offices.*

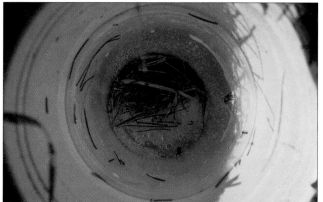

Pitfall trap in turf with two billbugs as well as other insects and mites.

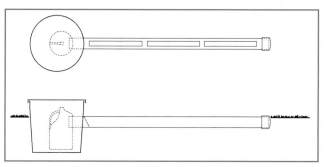

Diagram of a linear pitfall trap used to capture mole cricket nymphs. Two-inch diameter PVC pipe has slits cut out and the pipe is buried at ground level. Captured mole crickets, billbugs, and other insects crawl to the openings move down the pipe and fall into a plastic jug fitted over it.

Pitfall Trap. The pitfall trap consists of a plastic cup with a funnel and collection cup inserted inside. A hole the size of the cup is made in the turfgrass and the cup placed in the hole so the lip is at the thatch-soil level. A 16 to 20 oz. plastic cup usually fits in the hole made by a standard golf course cup cutter. Alcohol (20 to 50%) or water is placed in the collection cup. Insects crawling through the turf fall into the cup, through the funnel and into the collection cup of alcohol. Trap contents should be emptied daily. Small shelters may be placed over the top of the trap to keep rain out. The pitfall trap can be used to monitor and detect the presence of chinch bugs, adult billbugs and many other surface inhabiting insects.

A **linear pitfall trap** has been used in research to determine mole cricket nymphal activity. This trap uses a 3 to 4 foot section of two-inch diameter PVC pipe. A one-inch-wide slit is cut down the length of the pipe, one end is capped with a PVC end plug and the other end is inserted into the top side of a 3 to 5 gallon plastic bucket. The bucket is buried in the ground so that the PVC pipe opening is at ground level. A one-gallon plastic jug is fitted over the end of the pipe, within the bucket. Mole cricket nymphs that encounter the pipe,

emerge to crawl over the obstruction and fall into the slit. They then travel down the pipe and fall into the jug where they can be counted, recorded, and removed. This type of trap is also more efficient in capturing billbug adults.

Aspirator. An aspirator is a simple device consisting of a test tube-like container with a rubber stopper and two metal tubes inserted into the stopper. It is commonly used by entomologists to selectively collect individual insects. Operation consists of placing the plastic tube extension in the mouth and drawing air through it to create a vacuum at the tip of the other tube. The tube tip is then placed near the insect which is then drawn into the test tube. A fine screen over the intake tube prevents any foreign material from entering the mouth when air is drawn through the device.

Sweep Net. The sweep net is a particularly valuable tool for <u>detecting aphids, other pests in turfgrass, and collecting butterflies</u>. The net frame should be sturdy (beating type frame) and the net bag made of a solid rather than of a fine mesh cloth. With a motion similar to that used when using a long handled broom to sweep off a sidewalk, the <u>net is swept back and forth across the turf so the rim strikes just the grass blades</u>. The net is then turned inside out and the contents emptied onto a white pan or cloth. If aphids are present, they will collect in the net. If chinch bugs are present, they will also be collected because of their habit of crawling up to the tips of the grass blades during the day. Sweep net samples taken at dusk in the spring and fall months can be used to detect clover and winter grain mites. Many other insects will also be collected when "sweeping" turfgrass, but most of these do no damage. <u>You will be surprised at what is collected using the sweep net over turf</u>. It can be an informative experience.

Using a sweep net to sample aphids on turf.

Hand Lens. The hand lens is ***a most useful tool*** for identifying insects, disease lesions, grass species and other elements of the turfgrass environment. Use of the lens also projects a professional image to those nearby. Generally, lenses of 10X are adequate for most purposes. Such lenses can be purchased for $10 or less.

Source: <u>Pitfall traps</u>, <u>aspirators</u> and <u>sweep nets</u> can be purchased from entomological specialty companies. <u>Hand lenses</u> are available from college book stores, coin and stamp stores, and biological supply companies listed in the appendices of this book.

A hand lens is NOT a magnifying glass. You must hold the lens firmly to your eye and bring the object to be examined up to the lens until in focus. <u>Do not move the lens to the object like a magnifying glass</u>.

Rating Grid Frame. A PVC or wooden frame can be constructed and strung with nylon cord into one foot squares. When this grid is placed on the turf, each square that contains evidence of insect activity or damage (e.g., mole cricket tunnels, ant mounds, billbug damage, etc.) is counted. The total or percent of damaged squares is used as a rating. The data obtained are used to map areas, decide whether or not control is needed, or whether a treatment was effective.

An aspirator is used to easily capture small insects or mites.

This commercially available aspirator consists of a suction tube fitted with a fine screen to prevent material from entering the mouth and a holding tube.

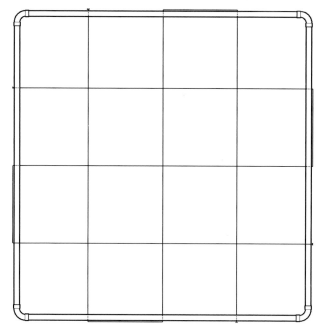
A four foot rating grid constructed using PVC pipe and nylon string that contains 16 squares.

Methods and Approaches

Flotation. Flotation is a method whereby water is used to detect the presence of insects such as the **chinch bug**. Cut both ends of a two-pound coffee can or similar container and remove the bottom off one end with tin snips to provide a sharp edge. The can is pushed thorough the turf into the soil, in an area suspected of being infested with chinch bugs, and water is added to the brim. If the water recedes, more should be added. If present, chinch bugs will float to the surface in 5 to 10 minutes. First and second instar chinch bugs are red-orange and tiny, so care must be taken not to miss them.

A one pound coffee can with bottom rim cut off (top left) **is pushed into turf and filled with water** *(top right)* **to float up chinch bugs** *(bottom).*

Soap Flush. Irritation is another effective method of detecting certain insects. Water, with a soap irritant added, is applied to turfgrass to irritate insects to the surface. This method is primarily effective in detecting caterpillar-type insects such as sod webworms and cutworms.

One readily available irritant is common household detergent. **Add two tablespoons of Joy® liquid detergent to two gallons of water** in a sprinkling can and apply the solution to one square yard of turfgrass where infestation is suspected. Cutworm and armyworm larvae usually surface within 10 minutes, but examination of the treated area 15 to 20 minutes after application is recommended, especially if sod webworm larvae are present. This method is especially useful for detecting cutworms and sod webworms on golf course greens. **Use before and/or after treatment will indicate treatment effectiveness**.

If the thatch is dry, irrigation before the test is advisable. This method does not normally bring soil-inhabiting insects such as grubs or billbug larvae to the surface, **but mole crickets readily emerge if they are near the turf surface**. Occasionally, a second application of the soapy water solution is needed to bring up mole crickets.

While in our experience this soap (Joy®) solution has not damaged turf, some occurrence of damage has been reported.

Mapping and Surveying -

Grubs in Lawns

Surveys should be conducted when grubs are in their second instar stage, when they are easy to find and identify, but generally too small to cause significant turf damage.

To survey, a **golf course cup cutter and hand lens are needed**, as well as the ability to identify the species of grub(s) found. Using graph paper, prepare a general map of the landscape or sports field indicating the location of buildings, major trees, shrubs, walkways, and other landmarks. For lawns, mark sampling spots that are approximately 6 to 10 foot apart in an approximate grid pattern. At each spot, take a sample and record the number of grubs found (be sure to record zeros!) on the map. Sports fields can be mapped and surveyed in a similar manner, only use a grid size of 10 to 20 feet.

Soap solution being applied to a one square yard area on a green to sample for cutworms.

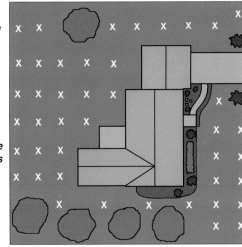

A general map of a home lawn showing places where samples are to be taken when surveying for grubs. No samples are needed in shaded turf since grub populations rarely reach damaging levels in such areas.

The standard 4.25-inch diameter cup cutter samples an area that is approximately 1/10 square foot, so multiplying the number of grubs per sample by 10 yields the average number per square foot. When 1 to 2 grubs per sample are found in several adjacent samples, that area of the lawn has a grub population of 10 to 20 grubs per square foot or more. For most lawns, when such a population is detected, sampling can be stopped since a "hot spot" has been identified.

Grubs on Golf Courses

Golf courses should be surveyed when grubs are in the second instar stage of development. When surveying should be done depends upon the species of grubs common to the site and location (consult your Pest Spectrum and Target Calendar for this information). Mid-August is about the time to survey for Japanese beetle and masked chafers in the northern states.

Each surveyor needs a **cup cutter**, **hand lens** and the **ability to identify the species of grub(s)** found in each sample. One person, equipped with an 8.5 X 11 inch sketch of the area(s) to be sampled (fairway and/or rough), records the number (including zeros, "0"s) and species of grubs reported at EACH SAMPLE POINT. A crew of two to four people can survey a standard 18 hole golf course in one day.

Sample Pattern - Starting about five feet in from the rough and following a **zigzag pattern**, a sample is taken and examined for grubs every 10 to 15 yards throughout the length of the fairway. Another approach is to line three or four surveyors across the width of the fairway and have each take a sample every 10 to 20 feet throughout the length of the fairway (the **transect pattern**).

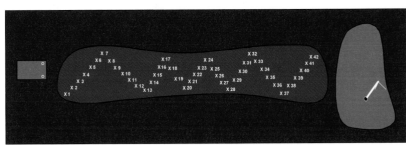

Zigzag pattern for sampling grubs on a golf course fairway.

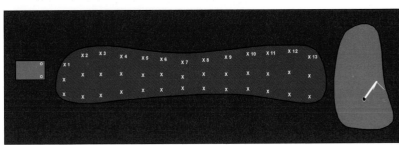

Transect pattern for sampling grubs on a golf course fairway.

Advice: Surveyors need not completely destroy a sample to determine if it has grubs. If the soil end of the sample is first split into halves and then quarters, the grubs which usually occur at the thatch-soil interface will be revealed. The objective is to expose the grubs but cause as little destruction of the sample as possible.

Care must be taken when replacing the sample in the sample hole. If the soil is dry, it may be necessary to add water to the hole before replacing the sample (bentgrass fairways are especially sensitive). When irrigation is available, the fairway should be irrigated as soon as possible after sampling to aid in sample survival.

After inspecting a core for grubs, carefully place it back into the hole, press it in until the top is level with the surrounding turf.

Irrigation of the turf after sampling will aid in turf recovery.

Mole Crickets on Golf Courses

Mapping spring mole cricket tunneling in turf has been proven to be a reliable method of identifying areas where the greatest risk of mole cricket damage will occur from August through November. As the first warm rain fronts pass through from March through May, overwintered adult mole crickets fly in mass to locate egg laying sites. These sites are usually moist and have thick turf and/or weed cover. Locations with considerable adult activity - tunneling and evidence of adults emerging and entering the turf - are places where adult females will lay most of their eggs.

Visual surveys and mapping should be performed in April to May (depending on the region of the country) to identify areas where spring adult mole cricket activity is most prevalent. Carefully inspect areas around greens, tees, bunkers, fairways and wetland sites (ponds, river banks, or marshy areas) and mark on a map of each fairway, areas where activity is heavy, moderate, light or not detectable. **Areas that qualify as heavy and moderate are prime sites for preventive controls.** Lightly infested and areas deemed not infested should be reinspected after egg hatch to detect any missed populations that may require curative treatments.

Heavy spring mole cricket tunneling usually indicates where turf loss will occur later in the season. Mark this area on your map!

Predicting Insect Pest Activity

Two methods have been used to help predict when a specific pest activity should occur - **Degree-Days** and **Plant Phenology**. It is important to remember that **both methods merely help predict WHEN a pest activity will occur but neither method can predict the MAGNITUDE of the activity or potential damage**.

Degree-Days - Since insect and mite pests are "cold blooded" animals, their development is slow at cool temperatures and faster at higher temperatures. The number of "heat units" above a baseline developmental threshold temperature an insect is exposed to each day (**=Degree-Days**), has a direct influence on how fast it will develop. **The base line developmental temperature for most insects is considered to be 50°F and air temperatures are often used**.

The **"daily average" method** is most commonly used to calculate the number of heat units above 50°F to which the pest is exposed. For example, if the low temperature for a day was 40°F and the high temperature was 70°F, the average temperature would be 55°F [(40 + 70)/2]. For this day 55° - 50° = 5 degree-day units. When these units are added to the previous days' units, one gets the **cumulative degree-day units**. (When calculation yields a negative number or zero, no degree-day units are accumulated - and zero is added for that day.) Once a certain number of degree-day units have been accumulated, pest activity can be expected. The following chart contains examples of degree-day accumulations for several turf pests.

Target Pest	Stage	Degree-Days*
Northern masked chafer	1st adults	898-905
"	90% adult flight	1377-1579
Southern masked chafer	1st adults	1000-1109
"	90% adult flight	1526-1679
Bluegrass billbug	1st adult activity	280-352
"	30% adult	560-624
"	70% egg hatch	925-1035
Hairy chinch bug	1st adult egg laying	198-252
"	1st egg hatch	522-702
Bluegrass webworm	1st gen. adult flight	864-900
"	2nd gen. adult flight	1900-2000
Larger sod webworm	1st gen. adult flight	846-882
"	2nd gen. adult flight	1980-2100
Cranberry girdler	peak adult flight	1080-1170

* using threshold base of 50°F and start date of February 1 in Ohio.

Phenological Indicators - *Apparently*, certain plants and insect pests develop at similar rates, and visible or measurable events in nature (= phenological occurrences) occur at the same time. The best phenological indicators are plants that have easily observed events (e.g., bud break, flowering - beginning, full bloom, or petal drop).

We recommend development of a **phenological record book**. Each year, **record in chronological order** the observable events seen and record pest activity at this time. After several years of keeping such records, it will be seen that certain plant events consistently occur in the same order and at the same time as certain pest events. **NOW, integrate these records into your Pest Spectrum and Target Calendar and you really have a useful tool!**

Forsythia in full bloom is the time that annual bluegrass weevil adults migrate to lay eggs.

Onset of full bloom in Vanhoutte spirea is the time that black turfgrass ataenius begin to lay eggs for the first generation.

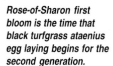

Rose-of-Sharon first bloom is the time that black turfgrass ataenius egg laying begins for the second generation.

Insect Specimen Identification

During the course of observations, surveys, monitoring and other activities associated with turf, tree and ornamental plant management, insects (pest and non-pest) whose identity is desired but unknown, are encountered. When identification is needed, sample specimens may be sent or delivered to the state Cooperative Extension entomologist at the state Land Grant University, or other professional entomologists with knowledge of insects on those hosts. Some state universities have a diagnostic clinic that provides such service.

Except for moths and butterflies, the following is suggested when submitting immature and adult stages of insects to an entomologist for identification:

1. Collect 5 to 10 live specimens. (be generous!)
2. Immediately after collection, boil a small quantity of water (a cup of water in a microwave oven is satisfactory) and drop in the specimens after the water reaches the boiling temperature. Wait 30 seconds and pour off the water.

3. <u>Drop</u> the "blanched" specimens <u>into</u> a leak-proof vial of <u>60 to 80% ethyl alcohol</u>. Vodka or denatured alcohol used as shellac thinner are sources of ethyl alcohol. DO NOT USE ISOPROPYL (rubbing) ALCOHOL.

4. <u>Record</u> the date, city, state and habitat location (e.g., leaves, thatch, soil, etc.) where the specimens were collected on paper and tape on the outside of the vial and/or written IN LEAD PENCIL in a small piece of paper and place inside the vial. (<u>Ballpoint pen ink will dissolve in the alcohol</u>.)

5. <u>Include a note</u> with the name, address and phone number of the inquirer, plus any additional information that might be helpful to the specialist.

6. <u>Ship</u> the vial, well protected, in a crush-proof container.

Moths such as cutworms or sod webworms and butterflies must be sent DRY and require special provisions to prevent destruction in shipping. Directions for such shipments should be obtained from the person or agency to whom the specimens are to be sent.

Developing a Reference Collection

When correctly identified, insect specimens can be assembled into a useful collection for <u>future reference</u> and <u>training</u> of employees. If properly preserved and sealed, the specimens in a collection last almost indefinitely.

Killing and Preservation. The larvae of insects are <u>soft-bodied and require special preparation</u> before long term preservation. Two methods may be used.

1. Live larvae are killed in a solution of one part 95% ethyl alcohol (shellac thinner) and one part xylene. (<u>Both ingredients are available from scientific supply companies</u>.) After 24 hours, transfer the specimens to a glass screw-top vial containing 75% ethyl alcohol for permanent preservation.

2. **OR**, drop live larvae into boiling water for 30-60 seconds. The specimens are then transferred to a glass screw-top vial containing 75% ethyl alcohol for permanent preservation. Replace the alcohol with clean 75% ethyl alcohol within 7 to 10 days.

Living adult beetles such as Japanese beetle, billbugs, masked chafers, etc., may be killed by direct immersion into 75% ethyl alcohol for at least 24 hours and transferred to clean 75% ethyl alcohol. Both larvae and adults of one species may be stored in one vial.

Caps for the screw-top vials used for permanent storage should have poly-seal™ inserts to prevent evaporation of the alcohol. Vials and caps may be purchased from the scientific and/or biological supply companies listed in Chapter 11.

Self-adhesive labels bearing the common name of the insect should be placed on the outside of each vial. ***Any label placed inside the vial must be written or printed using permanent, waterproof black India ink.*** Other inks will dissolve in the alcohol. Printed labels reduced and copied on most photocopy machines are alcohol safe. If in doubt, make some labels and place them in some 75% alcohol for 24 hours.

Display. Vials with specimens can be displayed in many ways. A wood 2" x 4" or 2" x 6" with evenly spaced holes of the same diameter as the vials works well. The holes and bottom fo the display may be lined with cork and sides with felt. Vial diameters are not always uniform.

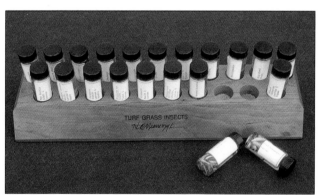

Example of a display rack made from 2" x 6" wood.

Notes

Notes

Chapter 10

Symptom Similarity

Symptoms caused by drought, improper or inadequate fertility, diseases, localized dry spots, winter desiccation, ball marks, and a multitude of other factors can be similar to those caused by insects and mites. Close examination (the **hands-and-knees method**) of the leaves, stem, crown, thatch <u>and underlying roots and soil</u> is necessary for proper diagnosis <u>and</u> determination of what (if any) action is needed. Anything less could lead to a "***costly assumption***."

In This Chapter

As a final effort to make this book relevant and useful to the reader, we present photos of various symptoms or conditions (some caused by insects and/or other factors) commonly seen on turf. Using the information presented in previous chapters of this book, and <u>your experience with turf</u>, examine each photo and determine the probable cause of the symptom(s). The actual cause of the symptom (as we saw it) is given on a following page.

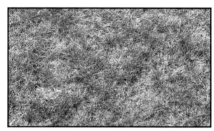

Symptom #1
Irregular white-yellow spots in <u>Kentucky bluegrass</u> appearing in late-June, early July.

Symptom #2
South facing fine <u>fescue-ryegrass</u> lawn has irregular patches turning brown in early July.

Symptom #3
<u>Kentucky bluegrass-ryegrass</u> lawn has circular patches turning brown in early July.

Symptom #4
Top dressing appears on surface of <u>bermudagrass</u> green in early-May, often C-shaped.

Symptom #5
Turf turns white next to building in April, <u>Kentucky bluegrass-perennial ryegrass</u> lawn.

Symptom #6
Irregular patches of <u>Kentucky bluegrass</u> turf dug up in mid-May.

Symptom #7
Medium to large wilt patches in <u>bentgrass</u> fairway turn gray-green in afternoon; early July.

Symptom #8
Medium to large patch of <u>bentgrass</u> turning gray-green and brown in mid-June.

Symptom #9
Circular areas of of <u>Kentucky bluegrass</u> under trees turn brown in late July.

Symptom #1 - Signs of <u>billbug</u> activity. Use the "tug test" to confirm billbug presence.

Symptom #2 - <u>Hairy chinch bug</u> damage, but could simply be summer drought-induced dormancy. Use "hands-and-knees" method to find chinch bugs. If not present, check soil moisture. Billbug larvae cause similar damage.

Symptom #3 - <u>Summer patch disease</u>. There are numerous diseases of lawns that can look like insect or environmental problems. There are several "patch diseases" that produce characteristic "frog-eyes" but their damage may appear as simple irregular spots.

Symptom #4 - <u>Overwintered sod webworm feeding</u> "cover." Peal back the top dressing cover to see if the particles are held together by silk.

Symptom #5 - Classic <u>clover mite damage</u>, but it could also be caused by winter grain mites or salts used for melting snow.

Symptom #6 - This is actually <u>damage by raccoons</u> feeding on earthworms! In wet conditions, they often pull up small patches of turf when they grasp the earthworms. This is commonly mistaken for animal foraging for white grubs.

Symptom #7 - Early signs of <u>black turfgrass ataenius grub damage</u>. Do not hesitate to make the proper diagnosis. If left untreated, the turf will die within days.

Symptom #8 - <u>Localized dry spot</u>. Black turfgrass ataenius grubs could cause the same type of symptom. Take a soil core to determine if soil is dry or grubs are present.

Symptom #9 - Though this may resemble green bug damage, it is <u>drought-induced dormancy</u> caused by competition between the turf and tree roots. Also, sample the turf to determine if billbug grubs are present, because they can also cause the turf to turn brown more rapidly where there is root competition.

Comparison Pairs of Symptoms

Symptoms 10 & 11 - The irregluar spots (left) were caused by female dog urine. They are often located in one section of the lawn or may be scattered over the lawn. We know of no pest that makes regularly spaced spots in lawns (right). After talking to the home owner, it was found that flower pots were placed on the lawn for a week before the plants were planted!

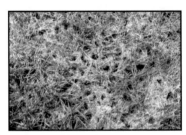

Symptom #10
Medium to large dead spots in <u>bluegrass/ryegrass</u> lawn, observed in mid-September.

Symptom #11
Regular, 6-inch bleached spots in <u>bluegrass/ryegrass</u> lawn suddendly appear in late June.

Symptoms 12 & 13 - Crows often pull up tufts of turf when eating grubs (left) while starlings feed on sod webworms in lawns (right). Recent studies in Ohio found that starlings feeding on greens in May and July are mainly feeding on black turfgrass ataenius adults! Those foraging in June, July, and August may be looking for black cutworm larvae, but they also pick up ants, ground beetles, sod webworms and a variety of other insects and invertebrates.

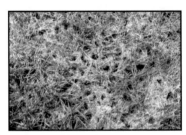

Symptom #12
<u>Crows</u> leaving tufts of turf on <u>bentgrass</u> green!

Symptom #13
Numerous small holes in <u>mixed cool-season turf</u> surface, appearing in mid-August.

Symptoms 14 & 15 - Mole activity (left) can be confused with pocket gopher digging and mound building (right). Moles can be found across North America while gophers are normally infest southern states. While moles "appreciate" white grubs and other insects, the <u>primary food of moles is earthworms</u>. Heavy activity, especially numerous straight tunnels, may merely be a sign of a healthy earthworm population. Pocket gophers feed on plant roots, tubers and bulbs.

Symptom #14
Large soil mounds pushed up <u>Kentucky bluegrass/ryegrass</u> turf in October in Ohio.

Symptom #15
Large soil mounds pushed up in <u>bluegrass/ryegrass</u> turf in October in California.

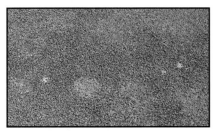

Symptom #16
Irregularly spaced sunken spots on bermuda-grass green, size of a quarter, August.

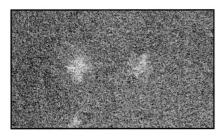

Symptom #17
Irregularly spaced spots of white bentgrass green, somewhat sunken, in mid-June.

Symptom #18
Bentgrass fairway wilts in September and pulls up like loose carpet.

Symptom #19
Small mounds of soil appear in late March in golf course Kentucky bluegrass rough.

Symptom #20
Small mounds of loose soil appear overnight on bermudagrass green in September.

Symptom #21
Two-inch diameter mound of soil appears in July on bermudagrass green in Arkansas.

Symptom #21
Bermudagrass in September appears thin and clumped, irrigation appears to miss spot.

Symptom #23
Large, 3/4-inch hole extends straight into soil, grass blades attached to edge with silk.

Symptom #24
Gray-brown larvae found in mass under one-inch diameter dead spots, March.

Symptom #25
Large irregular patches of dead bermuda-grass in lawn in May.

Symptom #26
Large irregular patches of cool-season lawn suddenly turn brown in July.

Symptom #27
Bermudagrass in South Florida has "ripple" pattern of brown and green in January.

Symptom #28
Irregular patches of Kentucky bluegrass in mid-July look droughted on bunker slope.

Symptom #29
Bermudagrass court in south Florida with large patches turning brown in December.

Symptom #30
Large area of Kentuckybluegrass/ryegrass lawn thins and browns in July.

Symptom #16 - Though these may look like cutworm spots, they are merely <u>ball marks</u>. To be sure, attempt to separate the turf to see if there is a burrow in the middle or to one side.

Symptom #17 - From a distance, <u>dollar spot symptoms</u> can look like cutworm feeding spots or billbug larval damage.

Symptom #18 - OK, so this is an easy one! <u>White grubs</u> ---- BUT which species?

Symptom #19 - <u>Green June beetle</u> overwintered grubs often clean out their burrows in early spring.

Symptom #20 - <u>Earwig females</u> often dig into sand-based greens in September through December in southern turf areas. These females are setting up brood chambers. Mowing clears off the sand.

Symptom #21 - <u>Several beetles may be found burrowing into greens in southern turf</u>. This is a mound from the *Geotrupes* beetle (earth-boring dung beetle). Little can be done to prohibit these periodic invaders.

Symptom #22 - <u>Bermudagrass mite damage</u> is often mistaken for drought stress, inconsistent irrigation coverage or other insect or disease problem. Their activity often causes bermudagrass to appear thin from a distance. Close inspection will reveal the typical witch's-brooming.

Symptom #23 - <u>Wolf spiders</u> commonly dig burrows in turf. Use a soap irritant solution to flush the spider from its burrow if you are unsure.

Symptom #24 - The <u>larvae of March flies</u> (gray-brown maggots with black head capsules) feed on winter killed patches of turf. They are often blamed for the damage, though diseases are usually the real culprit.

Symptom #25 - <u>Spring dead spot disease</u> of bermudagrass. Damage is very similar to hunting or Phoenician billbug damage.

Symptom #26 - <u>Roughstalked bluegrass</u> is a common perennial grass weed in cool-season lawns. This grass rapitly enters dormancy in hot or dry weather conditions. When this occurs in July, it may look like billbug or chinch bug damage. Dormancy in September looks like grub damage.

Symptom #27 - <u>Light frost damage</u> to bermudagrass that has not entered dormancy can cause discoloration that looks like a variety of insect, mite, or disease symptoms.

Symptom #28 - This is <u>bluegrass billbug damage</u> which is commonly mistaken for drought stress or disease.

Symptom #29 - <u>Hunting billbug damage</u> is often mistaken for bermudagrass winter dormancy, spring dead spot and delayed spring greenup.

Symptom #30 - <u>Leafspot disease (melting out)</u> in lawns can resemble a variety of damaging insect symptoms - billbug, chinch bug, sod webworm, white grubs.

We trust that these illustrations demonstrate that diagnosis "from afar" can lead to misdiagnosis. The "hands-and-knees" method (and this book!) is far more likely to lead to proper diagnosis.

Chapter 11

Appendix

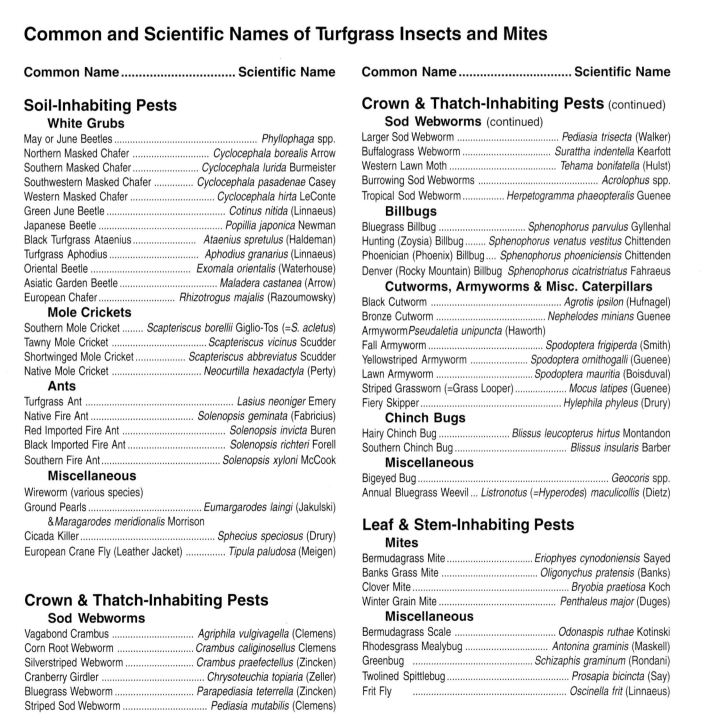

Common and Scientific Names of Turfgrass Insects and Mites

Common Name .... Scientific Name

Soil-Inhabiting Pests
White Grubs
May or June Beetles .... *Phyllophaga* spp.
Northern Masked Chafer .... *Cyclocephala borealis* Arrow
Southern Masked Chafer .... *Cyclocephala lurida* Burmeister
Southwestern Masked Chafer .... *Cyclocephala pasadenae* Casey
Western Masked Chafer .... *Cyclocephala hirta* LeConte
Green June Beetle .... *Cotinus nitida* (Linnaeus)
Japanese Beetle .... *Popillia japonica* Newman
Black Turfgrass Ataenius .... *Ataenius spretulus* (Haldeman)
Turfgrass Aphodius .... *Aphodius granarius* (Linnaeus)
Oriental Beetle .... *Exomala orientalis* (Waterhouse)
Asiatic Garden Beetle .... *Maladera castanea* (Arrow)
European Chafer .... *Rhizotrogus majalis* (Razoumowsky)

Mole Crickets
Southern Mole Cricket .... *Scapteriscus borellii* Giglio-Tos (=*S. acletus*)
Tawny Mole Cricket .... *Scapteriscus vicinus* Scudder
Shortwinged Mole Cricket .... *Scapteriscus abbreviatus* Scudder
Native Mole Cricket .... *Neocurtilla hexadactyla* (Perty)

Ants
Turfgrass Ant .... *Lasius neoniger* Emery
Native Fire Ant .... *Solenopsis geminata* (Fabricius)
Red Imported Fire Ant .... *Solenopsis invicta* Buren
Black Imported Fire Ant .... *Solenopsis richteri* Forell
Southern Fire Ant .... *Solenopsis xyloni* McCook

Miscellaneous
Wireworm (various species)
Ground Pearls .... *Eumargarodes laingi* (Jakulski)
&*Maragarodes meridionalis* Morrison
Cicada Killer .... *Sphecius speciosus* (Drury)
European Crane Fly (Leather Jacket) .... *Tipula paludosa* (Meigen)

Crown & Thatch-Inhabiting Pests
Sod Webworms
Vagabond Crambus .... *Agriphila vulgivagella* (Clemens)
Corn Root Webworm .... *Crambus caliginosellus* Clemens
Silverstriped Webworm .... *Crambus praefectellus* (Zincken)
Cranberry Girdler .... *Chrysoteuchia topiaria* (Zeller)
Bluegrass Webworm .... *Parapediasia teterrella* (Zincken)
Striped Sod Webworm .... *Pediasia mutabilis* (Clemens)

Common Name .... Scientific Name

Crown & Thatch-Inhabiting Pests (continued)
Sod Webworms (continued)
Larger Sod Webworm .... *Pediasia trisecta* (Walker)
Buffalograss Webworm .... *Surattha indentella* Kearfott
Western Lawn Moth .... *Tehama bonifatella* (Hulst)
Burrowing Sod Webworms .... *Acrolophus* spp.
Tropical Sod Webworm .... *Herpetogramma phaeopteralis* Guenee

Billbugs
Bluegrass Billbug .... *Sphenophorus parvulus* Gyllenhal
Hunting (Zoysia) Billbug .... *Sphenophorus venatus vestitus* Chittenden
Phoenician (Phoenix) Billbug .... *Sphenophorus phoeniciensis* Chittenden
Denver (Rocky Mountain) Billbug *Sphenophorus cicatristriatus* Fahraeus

Cutworms, Armyworms & Misc. Caterpillars
Black Cutworm .... *Agrotis ipsilon* (Hufnagel)
Bronze Cutworm .... *Nephelodes minians* Guenee
Armyworm*Pseudaletia unipuncta* (Haworth)
Fall Armyworm .... *Spodoptera frigiperda* (Smith)
Yellowstriped Armyworm .... *Spodoptera ornithogalli* (Guenee)
Lawn Armyworm .... *Spodoptera mauritia* (Boisduval)
Striped Grassworn (=Grass Looper) .... *Mocus latipes* (Guenee)
Fiery Skipper .... *Hylephila phyleus* (Drury)

Chinch Bugs
Hairy Chinch Bug .... *Blissus leucopterus hirtus* Montandon
Southern Chinch Bug .... *Blissus insularis* Barber

Miscellaneous
Bigeyed Bug .... *Geocoris* spp.
Annual Bluegrass Weevil ... *Listronotus* (=*Hyperodes*) *maculicollis* (Dietz)

Leaf & Stem-Inhabiting Pests
Mites
Bermudagrass Mite .... *Eriophyes cynodoniensis* Sayed
Banks Grass Mite .... *Oligonychus pratensis* (Banks)
Clover Mite .... *Bryobia praetiosa* Koch
Winter Grain Mite .... *Penthaleus major* (Duges)

Miscellaneous
Bermudagrass Scale .... *Odonaspis ruthae* Kotinski
Rhodesgrass Mealybug .... *Antonina graminis* (Maskell)
Greenbug .... *Schizaphis graminum* (Rondani)
Twolined Spittlebug .... *Prosapia bicincta* (Say)
Frit Fly .... *Oscinella frit* (Linnaeus)

Entomological Terms

Incomplete or gradual metamorphosis [life cycle] (example - hairy chinch bug)

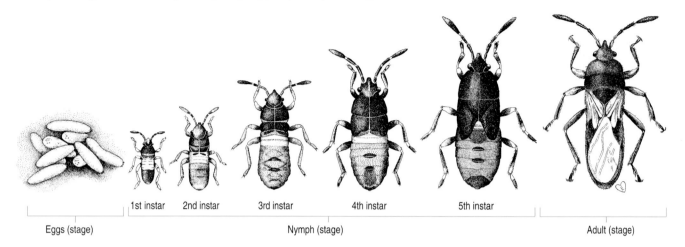

| 1st instar | 2nd instar | 3rd instar | 4th instar | 5th instar |

Eggs (stage)   Nymph (stage)   Adult (stage)

Complete metamorphosis [life cycle] (example - May/June beetle)

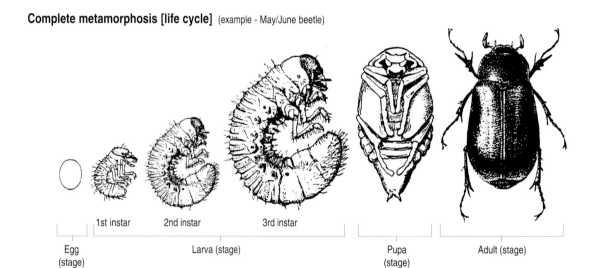

| 1st instar | 2nd instar | 3rd instar |

Egg
(stage)   Larva (stage)   Pupa
(stage)   Adult (stage)

Insect Morphology Terms

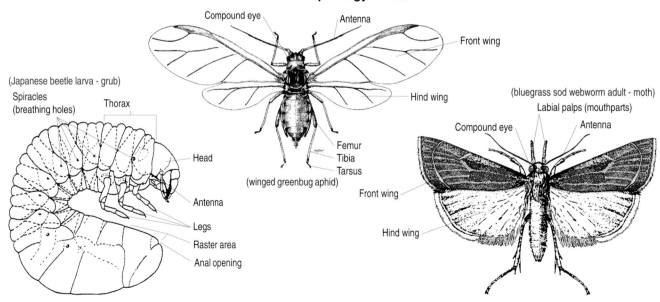

(Japanese beetle larva - grub)
Spiracles
(breathing holes)
Thorax
Head
Antenna
Legs
Raster area
Anal opening

Compound eye
Antenna
Front wing
Hind wing
Femur
Tibia
Tarsus
(winged greenbug aphid)

(bluegrass sod webworm adult - moth)
Labial palps (mouthparts)
Compound eye
Antenna
Front wing
Hind wing

Insect Morphology Terms

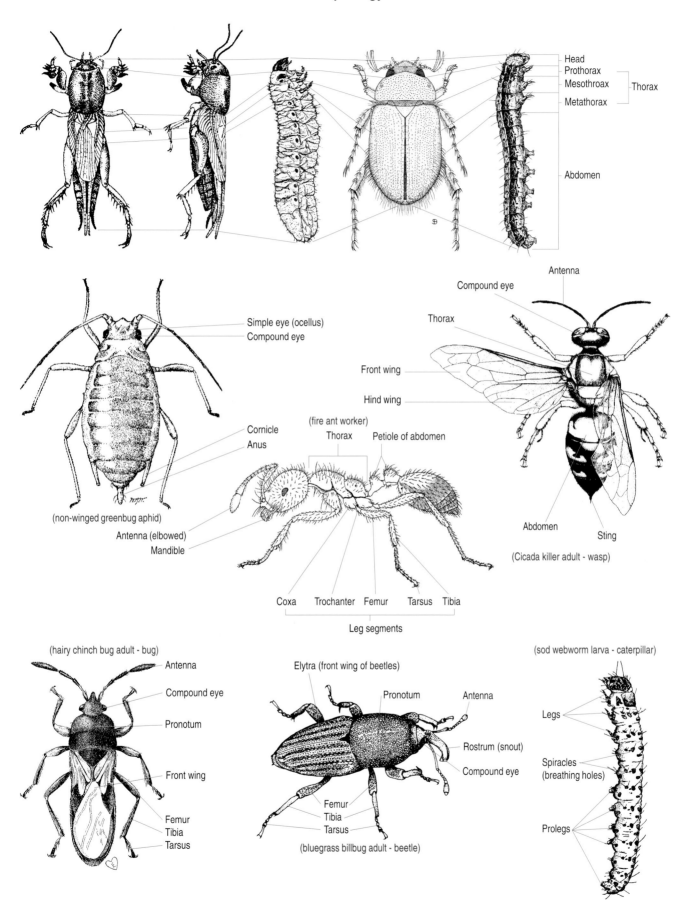

Head
Prothorax
Mesothroax
Metathorax
Thorax
Abdomen

Simple eye (ocellus)
Compound eye

Antenna
Compound eye
Thorax

Front wing
Hind wing

Cornicle
Anus

(fire ant worker)
Thorax
Petiole of abdomen

(non-winged greenbug aphid)

Antenna (elbowed)
Mandible

Abdomen
Sting

(Cicada killer adult - wasp)

Coxa Trochanter Femur Tarsus Tibia

Leg segments

(hairy chinch bug adult - bug)

Antenna
Compound eye
Pronotum

Front wing

Femur
Tibia
Tarsus

Elytra (front wing of beetles)
Pronotum
Antenna

Rostrum (snout)
Compound eye

Femur
Tibia
Tarsus

(bluegrass billbug adult - beetle)

(sod webworm larva - caterpillar)

Legs

Spiracles
(breathing holes)

Prolegs

141

References

Books of General Interest

A Field Guide to Insects (of America North of Mexico). D.J. Borror & R.E. White (1970), The Peterson Field Guide Series. Houghton Mifflin Company, Boston, MA. 404 pp.

Handbook of Integrated Pest Management for Turf and Ornamentals. A.R. Leslie, ed. (1994), Lewis Publishers, Boca Raton, FL.

Insect Pests. G. S. Fichter. Golden Press, Inc., NY, NY.

Introduction to Insect Pest Management. R.L. Metcalf & W. Luckmann (1975), John Wiley & Sons, NY. 587 pp.

Insects that Feed on Trees and Shrubs. W.T. Johnson & H.H. Lyon (1976), Cornell Univ. Press, Ithaca, NY. 464 pp.

IPM Handbook for Golf Courses. G. Schumann, P. Vittum, M. Elloit, & P. Cobb (1997). Ann Arbor Press, Inc., Chelsea, MI.

Practical Turfgrass Management. J.H. Madison (1971), Van Norstrand Reinhold Co., N. Y. 466 pp.

Lawn Care. H.F. Decker & J.M. Decker (1988), Prentice-Hall, Englewood Cliffs, NJ.

The Lawn Book. N.L. Wise (1961), Bowen Press, Inc., Decatur, GA. 250 pp.

Turfgrass Science and Culture. J.B. Beard (1973), Prentice Hall, Inc., Englewood Cliffs, N J. 658 pp.

Turfgrass Bibliography 1672 to 1972. J.B. Beard, H. S. Beard & D.P. Martin (1977), Mich. State Univ. Press. 730 pp.

Turfgrass Science. A.A. Hanson & F.V. Juska (1969), Amer. Soc. Agronomy, Madison, WI. 715 pp.

Books on Turfgrass Insects

Advances in Turfgrass Entomology. H.D. Niemczyk and B.G. Joyner (1982), Hammer Graphics, Piqua, OH. 150 pp.

Destructive Turf Insects. H.D. Niemczyk (1981), HDN Books, Wooster, OH. 48 pp.

Destructive Turfgrass Insects, Biology, Diagnosis, and Control. D.A. Potter (1998), Ann Arbor Press, Inc., Chelsea, MI. 344 pp.

Managing Turfgrass Pests. T.L. Watschke, Peter H. Dernoeden, and David J. Shetlar (1995), Lewis Publishers, CRC Press, Inc., 2000 Corporate Blvd. N.W., Boca Raton, FL. 361 pp.

Scotts Guide to the Identification of Turfgrass Diseases and Insects. The O.M. Scott and Sons Company (1987), Marysville, OH. 105 pp.

Turfgrass Insect and Mite Manual, 3rd Edition. D.J. Shetlar, P.R. Heller, and P.D. Irish (1983), The Pennsylvania Turfgrass Council, Inc., Bellefonte, PA. 63 pp.

Turfgrass Insects of the United States and Canada, 2nd Edition. P. J. Vittum, M.G. Villani, & H. Tashiro (1999), Cornell University Press, Ithaca, NY. 422 pp.

Handbook of Turfgrass Insect Pests. R.L. Brandenburg & M.B. Villani (ed.)(1995), Entomological Society of America, Lanham, MD. 140 pp.

Articles on Thatch and Accelerated Degradation

Niemczyk, H. D. 1977. Thatch: a barrier to control of soil-inhabiting pests of turf. Weeds, Trees and Turf. 16 (2): 16-19.

Niemczyk, H. D. 1993. Accelerated degradation of turfgrass pesticides: a review. International Turfgrass Soc. Res. J. 7: 148-151.

Articles on Specific Pests

Cameron, R.S. & N.E. Johnson. 1971. Biology and control of turfgrass weevil, a species of *Hyperodes.* New York State Coll. Agr. Ext. Bull. 1226. 8 p.

Cobb, P.P. & T.P. Mack. 1989. A rating system for evaluating tawny mole cricket, *Scapteriscus vicinus* Scudder, damage (Orthoptera: Gryllotalpidae). J. Entomol. Sci. 24: 142-144.

Johnson-Cicalese, J.M., G.W. Wolfe & C.R. Funk. 1990. Biology, distribution and taxonomy of billbug turf pests (Coleoptera: Crculionidae). Environ. Entomol. 19: 1037-1046.

Kennedy, M.K. 1980. "New" webworm pests in Michigan lawns. Amer. Lawn Applicator 1(4): 11-14.

Kennedy, M.K. 1981. Chinch bugs: biology and control. Amer. Lawn Applicator 2(4): 12-15.

Fleming, W.E. 1972. Biology of the Japanese beetle. USDA Tech. Bull. 1449. 129 pp.

Niemczyk, H.D. 1976. *Ataenius spretulus:* a new grub problem in golf course turf. Golf Supt. 44: 26-29.

Niemczyk, H.D. and G.W. Wegner. 1979. Controlling the black turfgrass ataenius. Golf Course Management 47(4): 29-37.

Niemczyk, H.D. 1978. The winter grain mite: A Winter Pest of Turf. Weeds, Trees and Turf. 17(2): 22-23.

Niemczyk, H.D. 1980. Proper pesticide application for effective greenbug control. Lawn Care Industry 4(2): 16-19.

Oliver, A.D. and K.N. Komblas. 1981. Part 1. Southern chinch bug in Louisiana. Amer. Lawn Applicator 2(4): 26-31. Part 2. 2(5): 12-16.

Potter, D.A. 1981. Seasonal emergence and flight of northern and southern masked chafers in relation to air and soil temperature and rainfall patterns. Environ. Entomol. 10: 793-797.

Reinert, J.A. 1974. Control of the southern chinch bug and sod webworm in Florida turfgrass-Effect of water rate and formulation of Dursban® insecticide. Down to Earth 29(4): 10-13.

Reinert, J.A. 1982. The Bermudagrass stunt mite. USGA Green Section Record. 20(6): 9-12.

Sparks, B. 1992. Controlling fire ants in urban areas. Univ. Ga. Coop. Ext. Bull. 1068.

Streu, H.T. & J.B. Gingrich. 1972. Seasonal activity of the winter grain mite in turfgrass in New Jersey. J. Econ. Entomol. 65: 427-430.

Tashiro, H. 1976. Hyperodes weevil: A serious menace to P. annua in Northeast. Golf Superintendent 44(3): 34-37.

Tashiro, H., et al. 1969. Biology of the European chafer, Amphimallon maialis (Coleoptera-Scarabaeidae) in northeastern United States. N. Y. State Agr. Exp. Sta. Bull. 828: 71 pp.

Tolley, M.P. & W.H. Robinson. 1986. Seasonal abundance and degree-day prediction of sod webworm (Lepidoptera: Pyralidae) adult emergence in Virgina. J. Econ. Entomol. 79: 400-404.

Tolley, M.P. & H.D. Niemczyk. 1988. Seasonal abundance, oviposition activity, and degree-day prediction of adult frit fly (Diptera: Chloropidae) occurrence on turfgrass in Ohio. Environ. Entomol. 17: 855-862.

Walker, T.J. (Ed.) 1984. Mole crickets in Florida. Florida Agr. Exp. Sta. Bull. 846: 54 pp.

Williamson, R.C. & D.J. Shetlar. 1995. Oviposition, egg location and diel periodicity of feeding by black cutworm (Lepidoptera: Noctuidae) on bentgrass maintained at golf course cutting heights. J. Econ. Entomo. 88: 1292-1295.

Vittum, P.J. & H. Tashiro. 1987. Seasonal activity of Listronotus maculicollis (Coleoptera: Curculionidae) on annual bluegrass. J. Econ. Entomol. 80: 773-778.

Pesticides

Apply Pesticides Correctly: A Guide for Commercial Applicators. USDA-EPA. Supt. Public Documents, Wash., DC. 26 p.

Recognition and Management of Pesticide Poisoning. U.S. Environ. Protection Agency, Office of Pesticide Programs, Washington, D. C. 20460 (Second Ed.) August 1977. EPAS-40 / 9-77-013. 75 pp.

Useful Trade Journals

Golf Course Management, 1421 Research Park Drive, Lawrence, KS 66044.

Greenmaster, Canadian Golf Course Superintendents Association, 2000 Weston Road, Suite 203, Weston, Ontario M9N IX3, Canada.

Grounds Maintenance, Intertec Publishing Corporation, 9800 Metcalf Avenue, Overland Park, KS 66212.

Landscape Management, Advanstar Communications, 7500 Old Oak Blvd., Middleburg Heights, OH 44130.

Lawn & Landscape Maintenance, 4012 Bridge Avenue, Cleveland, OH 44113.

Southern Golf, Landscape & Resort Management, Brantwood Publications, Inc., Northern Plaza Station, Clearwater, FL 34621.

Sports Turf, An Adams Publishing Co., 68-860 Perez Rd., Suite J, Cathedral City, CA 92234.

Turf & Landscape Press, ARGUS Agronomics, P. 0. Box 1420, Clarksdale, MS 38614.

Turf, NEF Publishing Company, P. .0. Box 391, 50 Bay Street, Johnsbury, VT 05819.

Golf Course News, United Publications, Inc., 38 Lafayette Street, P. 0. Box 997, Yarmouth, ME 04096.

Turf & Recreation, Turf & Recreation Publishing Company, 123B King Street, Delhi, Ontario, N4B IX9, Canada.

Turf Craft Australia, Turfcraft, P. 0. Box 160, 200 Rouse Street, Port Melbourne, Victoria, 3207, Australia.

Slide Sets of Turfgrass Insects

PLCAA, 1225 Johnson Ferry Rd., Marietta, GA 30067.

New York State Turfgrass Association, 210 Cartwright Blvd., Massapequa, NY 17762.

GCSAA, 1421 Research Park Drive, Lawrence, KS 66044.

University of Florida, Cooperative Extension Service, Gainesville, FL 32611.

Sources of Entomological Equipment

American Biological Supply Company
288 B-1 East Green St.
Westminster, MD 21157
(410) 876-8599

BioQuip Products
17803 LaSalle Ave.
Gardena, CA 90248-3602
(310-324-0620)
www.bioquip.com

Ben Meadows Company
3589 Broad Street
Atlanta, GA 30341
(800-241-6401)
www.benmeadows.com

Carolina Biological Supply Company
2700 York Road
Burlington, NC 27215
800-334-5551
www.carolina.com

Connecticut Valley Biological Supply Company
P.O. Box 326
82 Valley Road
Southampton, MA 01073
(800-628-7748)

Fisher Scientific
711 Forbes Avenue
Pittsburgh, PA 15219-4785
(800-766-7000)
www.fishersci.com

Forestry Supplier, Inc.
P.O. Box 8397
Jackson, MS 39284-8397
(800-647-5368)
www.forestry-suppliers.com

Gempler's
100 Countryside Drive
P.O. Box 270
Belleville, WI 53508
(800-382-8473)
www.gemplers.com

NASCO Science
901 Janesville Ave.
Fort Atkinson, WI 53538-0901
(920-563-2446)
www.nascofa.com

WARD'S
5100 West Henrietta Road
P.O. Box 92912
Rochester, NY 14692-9012
(800-962-2660)
www.wardsci.com

Information About Turf Insecticides / Miticides

Pesticide (Common Chemical Name)	Common Market Name(s)	Classification	Oral LD$_{50}$[1] (mg/kg)[2]	Dermal LD$_{50}$[1] (mg/kg)[2]	Manufacturer
acephate	Orthene	organophosphate	980	10250	Valent, Chevron, Micro Flo
azadiractin (=neem, azatin)	Neem, Turplex	botanical	>5000	>2000	Grace-Sierra, The Scotts Co.
Bacillus thuringiensis	BT, Bactospeine, Dipel, Javelin, Thuricide Vectobac, Condor T/O, & others	spores + crystalline delta-endotoxin, microbial	none	none	Numerous - Abbott DuPont, Upjohn Co Ecogen, etc.
Beauveria bassiana	White fungus, *Beauveria*	fungus	none	none	Troy Biosciences
bendiocarb[3]	Ficam, Turcam	carbamate	156	>1000	AgrEvo (Avantis)
bifenthrin	Talstar	pyrethroid	375	>2000	FMC
carbaryl	Carbaryl, Sevin	carbamate	246	>4000	Rhone-Poulenc, Drexel
chlorpyrifos	Dursban, Pageant	organophosphate	270	2000	DowAgroSciences
cyfluthrin	Decathlon, Tempo	pyrethroid	826	>2000	Bayer, Olympic
deltamethrin	DeltaGard	pyrethroid	128	>2000	Aventis
diazinon	Diazinon, Spectracide	organophosphate	400	3600	Novartis, Mico Flo, Drexel
dicofol	Docofol, Kelthane	chlorinated hydrocarbon	595	>5000	Rohm & Haas
ethoprop	Mocap	organophosphate	61.5	2.4	Aventis
fenoxycarb	Award, Logic	carbamate	16800	>5000	Novartis
fipronil	Chipco Choice,	phenyl pyrazole	97	>2000	Aventis
fluvalinate	Mavrik Aqua Flow	pyrethroid	282	>2000	Zoecon
fonofos[3]	Crusade, Dyfonate	organophosphate	8-17.5	25	Zeneca
halofenozide	MACH2	diacylhydrazine	2850	>2000	Rohmid
hydramethylnon	Amdro, Maxforce	unclassified	1300	>5000	Cyanamid
imidacloprid	Merit	chloronicotinyl	450	>2000	Bayer, Olympic
isazofos[3]	Triumph	organophosphate	40-60	118	Novartis
isofenphos[3]	Oftanol	organophosphate	20	700	Bayer
lambda-cyhalothrin	Scimitar	pyrethroid	79	632	Zenica
malathion	Cythion, Malathion	organophosphate	1000	4100	Hopkins, Drexel
metaldehyde	Bug-Geta, Deadline, Slug-Geta	metacetaldehyde	283	—	Micro Flo, Pace International
permethrin	Astro, Ambush, Pounce	pyrethroid	450-4000	>4000	FMC, Zeneca, AgrEvo
spinosyn	Conserve	spinosad	>2000	>2000	Dow AgroSciences
thiamethoxam[4]	Meridian	neonicotinoid	1,563	>2000	Novartis
trichlorfon	Dylox, Proxol	organophosphate	250	> 2100	Bayer, Aventis

[1]Farm Chemicals Handbook '98 (Meister Publishing Co., Willoughby, OH), and technical data information where available.
[2]Equals milligrams per kilogram of body weight applied orally or dermally. (1 milligram = 1/1,000 of a gram, 454 grams = 1 lb.)
[3]Product manufacturing currently withdrawn (as of May, 2000) in United States.
[4]Commercial product not available at the writing of this book, but expected in near future.

Notes

Notes

Notes